Animal Tales
from a 25-year Zoo Safari

Postcards from the ZOO

DARILL CLEMENTS

WITH PHOTOGRAPHY BY RICK STEVENS

HarperCollins*Publishers*

A portion of the proceeds of this book will be donated to the Taronga Foundation to assist the Zoos' animal behavioural enrichment programs.

While every care has been taken to trace and acknowledge copyright material, the author would be grateful to hear of any ommissions. She can be contacted care of the publisher.

HarperCollins*Publishers*

First published in Australia in 2002
by HarperCollins*Publishers* Pty Limited
ABN 36 009 913 517
A member of the HarperCollins*Publishers* (Australia) Pty Limited Group
www.harpercollins.com.au

Copyright © Darill Clements 2002
Photography by Rick Stevens copyright © Rick Stevens

The right of Darill Clements to be identified as the moral rights author
of this work has been asserted by her in accordance with the *Copyright Amendment
(Moral Rights) Act 2000* (Cth).

This book is copyright.
Apart from any fair dealing for the purposes of private study, research,
criticism or review, as permitted under the Copyright Act, no part may be
reproduced by any process without written permission.
Inquiries should be addressed to the publishers.

HarperCollins*Publishers*
25 Ryde Road, Pymble, Sydney, NSW 2073, Australia
31 View Road, Glenfield, Auckland 10, New Zealand
77–85 Fulham Palace Road, London, W6 8JB, United Kingdom
Hazelton Lanes, 55 Avenue Road, Suite 2900, Toronto, Ontario M5R 3L2
and 1995 Markham Road, Scarborough, Ontario M1B 5M8, Canada
10 East 53rd Street, New York NY 10022, USA

National Library of Australia Cataloguing-in-Publication data:

Clements, Darill.
    Postcards from the zoo.
    ISBN 0 7322 7239 4.
    1. Taronga Zoo (Sydney, N.S.W.) – Anecdotes.
    2. Zoo keepers – New South Wales – Sydney – Anecdotes.
    3. Clements, Darill.
    I. Stevens, Rick (Kenneth). II. Title
590.739441

Cover photography by Rick Stevens and Taronga Zoo
Cover and internal design by Melanie Calabretta, HarperCollins Design Studio
Typeset by HarperCollins in 11/14.5 Sabon
Printed and bound in Australia by Griffin Press on 79gsm Bulky Paperback White

5 4 3 2 1   02 03 04 05

Dedicated to
the animals of Taronga and
Western Plains Zoos — present, past
and future — and the photographers
who immortalise their wild beauty.

In memory of
my father, mother and sister.

Contents

Foreword
My Zoo Opening and Closing Times
vii

Chapter 1
Apes – Human and Hairy
1

Chapter 2
Some of my Friends are Horny
41

Chapter 3
Tales of a Monkey or Two
60

Chapter 4
Big Cats and Small
74

CHAPTER 5
Dances with Brolgas
101

CHAPTER 6
Why Do You Love Koalas?
121

CHAPTER 7
85 Years Caring for Wildlife
137

CHAPTER 8
Pandamonium
163

CHAPTER 9
Elephants I'll Never Forget
178

CHAPTER 10
All Creatures Great and Tall
198

ACKNOWLEDGMENTS
222

ABOUT THE AUTHOR AND PHOTOGRAPHER
225

Foreword

My Zoo Opening and Closing Times

At 2.00 p.m. on 1 December 2000 I was 'kidnapped' by a group of grown men dressed in animal suits. They looked pretty silly as they carried me from my office into the grounds of beautiful Taronga Zoo, situated on the shores of Sydney Harbour, but I looked even sillier when they put me in a large seal transport crate on the back of a truck and drove me around the Zoo for an hour.

It was a busy Friday afternoon and Zoo visitors of all ages laughed, pointed and looked quizzically at the poor unfortunate little woman in the crate with the very embarrassed look on her face. Perhaps they thought it was a regular Friday afternoon parade at Taronga. Zoo staff enjoyed it, too, and came out from everywhere to poke me with sticks and two-way radio aerials and generally make fun of me.

So ended my privileged Zoo life and the days shared with my Zoo 'family' in what I consider to be the best public relations job in Sydney.

Over a quarter of a century before I was at a point in my life where I needed a change in direction. I had spent my youth dancing with the discipline, passion and focus which classical ballet demands. When I left school I danced and taught ballet and at the same time went to college to do a secretarial course, which my father always said would enable me to get a 'real' job one day. In those days the only avenue for a professional dance career was with The Australian Ballet and those positions were extremely hard to come by and way out of my reach. So to earn some 'real' money my first job saw me working for an insurance company briefly and after that I worked in an architectural practice for some years.

I still attended classes and taught ballet at night and on weekends and loved performing in a variety of shows, concerts and eisteddfods. But when the *grand jetés* were getting less *grand* and my *pirouettes* (and the architectural industry!) were slowing down somewhat, I decided to try something different.

In May 1975, at the age of 30, I answered an advertisement in the *Sydney Morning Herald* 'Positions Vacant' section for a public relations assistant at Taronga Zoo, Sydney. My husband, Rob, had pointed out the ad to me just as a joke. I had no interest in animals whatsoever — I had never even owned a pet. In fact, I walked on the other side of the street to avoid dogs because after being bitten by one as a child, they made me extremely nervous. So I surprised myself as well as my family when I showed genuine interest in the advertised Zoo job.

I went along to the job interview at Taronga Zoo and I very vividly remember being asked: 'You are not an animal "nutter" are you? Because if you do get the job you'd be here to work, not to play with the animals all day.' I remember thinking at the time this was a strange statement coming from a zoo person, because I assumed you should automatically love animals if you worked in such a place. (I later went on to understand that it can be difficult, dirty, frustrating and, at times, sad working in a zoo. But it was always exciting, often exhilarating and very, very satisfying. What you got in return went far beyond any ordinary job.) My response about not being very interested in animals must have done the trick for some inexplicable reason. The next thing I knew I got the job and was doing a *pas de chat* through the famous front entrance gates of Taronga Zoo at Mosman, feeling a little apprehensive and certainly having no idea of the amazing journey which lay ahead of me.

I often wondered over the next quarter of a century whether my father, who had died when I was 25 years old, would have considered my public relations position at the Zoo a 'real' job. I had a little five-year-old friend who certainly did not. As Yashwant Sinha began her first school holidays I enquired what she was going to be doing in her vacation. She listed several exciting activities and asked me what I was going to do in *my* school holidays. I explained that when you work you don't get school holidays. Yashwant looked at me in absolute amazement and said: 'But you don't work, you go to the Zoo every day!'

There were many times when I recalled this innocent observation, particularly on the nights I was at the Zoo until 11.00 p.m. doing night tours or meeting impossible deadlines; or on the cold winter mornings when I had to start at 4.00 a.m. because a radio station or television network needed access to the Zoo grounds to broadcast their breakfast show live.

Almost immediately after I joined the Zoo I began to rescue stray cats in its car parks, stopped to check pouches of animal road accident victims, planted native shrubs and installed bird baths in my garden, carried huge caterpillars to the local park instead of squashing them to prevent them eating my husband's favourite ornamental grape vine, read every natural history book I could get my hands on, recycled things to within a inch of their lives and stopped eating red meat.

After only three months I realised I had become an animal 'nutter', besotted with and inspired by the joy of nature, which has enriched my life immeasurably ever since. I have probably driven my real family mad and for that I apologise.

It was wonderful to be able to incorporate my lifelong love of music, art and performance into Zoo activities, too. I found that concerts of many different kinds fitted well into the Taronga culture and invariably enhanced visitors' appreciation of the animals. Music and animals cross many cultural boundaries. And to see the immediate results of my efforts on television, in print or hear the radio interviews I organised was also very satisfying. And to know that, in some small way, I and the many hard-working public relations officers I had the pleasure of working with over the years were making a difference for wildlife, the environment and this State's two magical zoological institutions.

As the years went on I was blessed with the warmth and genuine closeness of my Zoo 'family', which, in times of personal trouble or sadness always sheltered me, and all others who work there, and genuinely and enthusiastically shared and revelled in the personal highs as well.

For my 25 Zoo years, it was the animals that were my teachers. It was always the animals that delighted, surprised, enlightened, energised, made me laugh and, of course, sometimes saddened me. They inspired me and gave me a fund of stories I ultimately needed to write down and share. I hope I do them justice and that

I can, in some way, return a little of what they gave to me over the years.

As you can imagine, during that quarter of a century I witnessed many unique and historical Zoo milestones. Some other Zoo animal events, however, such as births, hatchings, releases, arrivals, escapes and transfers I saw repeated time and again. In recording some of these here I have sometimes combined details of several similar stories. I never tired of these events and adventures and found it was these magical animal stories that people wanted to hear more than anything else. The anecdotes, tall tales and true about the Zoos' inhabitants seemed to capture everyone's imagination. After all, this, to me, is what zoos are all about. They can provide 'magic moments' that have a positive impact on visitors of all ages which encourage and strengthen our empathy with the natural world.

There was always news, too, of Zoo financial joys or woes, political announcements, occasional industrial unrest, myriad special events, the odd dispute with Zoo neighbours over traffic or noise, sponsorship launches, competitions for visitors and countless new exhibits. All this news paled in comparison with the interest shown specifically in what the animals were up to. And so it should, because without the animals there wouldn't be zoos.

Taronga and Western Plains Zoos are like giant, living reference libraries for artists, sculptors, actors, writers and photographers. Every day I would see visitors sketching, photographing and observing the animals, as well as the trees, plants and flowers. It is the Zoos' mission to increase people's understanding of and empathy with all animals and their environments, and photography is a medium that certainly does this in a very graphic way.

So it is, now, with great fondness, a fair dose of anthropomorphism and with elephant-sized gratitude to the animals, and to my wonderful former zoo colleagues and my friends in the local and international media, that I recall my days at Taronga and Western Plains Zoos, from my diaries and from memory, and send these postcards as a record of my 25-year 'safari' there.

I could have chosen to write this book at least ten different ways but I have let the Zoo animals, and a few of the human animals, dictate the content and style. I'll leave it to others far better qualified than I to do something more historical, educational and scientific about New South Wales's two great zoological institutions.

How fortunate I am to be sending these picture postcards from the Zoo in concert with the *Sydney Morning Herald's* master wildlife photographer, and my long-time friend, Rick Stevens.

Rick came to love Taronga and Western Plains Zoos as much as I do and gave of his time and exceptional photographic talents freely to the Zoos. He is a master photographer and I valued his and the *Herald*'s support throughout my Zoo days. They both always treated me very kindly.

February 2002

Chapter One

Apes – Human and Hairy

For over 25 years the human animals in the Zoo never ceased to amaze me. I learned so much from Zoo staff and visitors in my very early days there, and it continued up until the very last moment in that crazy place I love so much.

I felt blessed that every day my work took me out into the beautiful gardens of Taronga. I would spend many hours out in the grounds seeking news stories, escorting media representatives, welcoming VIPs or liaising with the keeping and horticultural staff. All the while, I loved hearing what visitors had to say about the animals and about the Zoo in general. It was always enlightening, encouraging, sometimes embarrassing and often very, very amusing.

The entry gate and food-service lines proved great listening points and I made sure I 'queued' as often as possible to listen to visitor comments and family conversations about their Taronga visit. I felt a little voyeuristic at times but it was a window into the mind of that very important Zoo animal, the visitor, which was invaluable in our constant endeavour to improve our Zoos for the animals and the people who visit them.

One thing became obvious to me very quickly — Zoo visitors were particularly fascinated by the variety of primates we displayed at Taronga. The name 'primate' comes from the Latin 'primus', which means first or foremost. The hairy primates — the apes and monkeys — had enormous appeal to the 'top of the tree' human primates. However, many people seemed to have great difficulty determining what was an ape and what was a monkey. Invariably I heard visitors saying, 'Look at that cheeky monkey' while pointing to a boisterous chimpanzee.

On behalf of all the apes in Taronga Zoo, who take great umbrage at being called monkeys, I should explain the most basic and obvious difference. Most monkeys possess tails but apes do not. Apes also possess arms that are longer than their legs, very flexible wrists and powerful barrel-shaped chests. Taronga Zoo's ape collection has long included the subtle and sensitive Western Lowland Gorilla, the rather raucous chimpanzee, the languid orang-utan and the acrobatic Mueller's Gibbon.

Taronga's apes and monkeys have always provided me with much laughter, joy and some sadness. In fact, some of my best Zoo moments have been as a result of encounters with these intriguing animals.

I am sorry to say that, despite my best intentions not to give them human attributes, I do find them extremely like people, but I guess that is why we all relate so well to these other primates — our 'hairy' relatives.

Taronga Zoo's and in fact Australia's first gorilla was a young male called King Kong. His name was meant to evoke

GORILLA FACTS

THE WESTERN LOWLAND GORILLA (*Gorilla gorilla gorilla*) is the largest living primate and, together with the chimpanzee, is our closest living relative, sharing about 98 per cent of its genetic material with humans. The gorilla is classified as an endangered species and is almost certainly facing extinction in the wild, although the Western Lowland Gorilla is the most numerous and has the widest distribution. These gentle and peaceful animals are suffering horrendously from habitat destruction and are also hunted for the pet and bush-meat trades. They are found in the African country of Gabon, where they live in dense rainforest which is hot and humid with year-round rainfall.

The gorillas live in tightknit family groups of up to ten animals and this group, or troupe, is led by a dominant male called a silverback. Adult males weigh about 180 to 200 kg and stand at around 1.7 m tall, while the adult females are smaller, weighing about 90 kg, and are a more diminutive 1.5 m in height. They are mostly vegetarian, feeding on the flowers, leaves and stems of more than 100 species of succulent plants, which also provide them with moisture. They are known to also eat termites, ants and soil, and to select some plants specifically for medicinal purposes.

in the public mind images of a gigantic beast, just like the one in the movie. A new exhibit called the Gorilla Villa was built for this animal, and he arrived at Taronga in 1959 as a five-year-old. Despite being nothing remotely like his publicity build-up, Taronga's King Kong soon became a huge favourite with Zoo visitors. And so began the long history of gorillas in Sydney's Zoo.

When I started work as the public relations assistant at Taronga Zoo in 1975, Buluman and Betsy were the resident Western Lowland Gorillas. Zoo Archives record that Buluman was a juvenile gorilla when he arrived at Taronga Zoo via the United States in 1961. It seems he was born in Cameroon in 1958, where he had been captured and then taken to the States by an animal trader before ending up at Taronga. Betsy was purchased from the same animal trader and arrived at Taronga in 1963 to team up with Buluman. I have to say that those were the bad old days, and acquiring animals in such a way has largely ceased for the 21st century zoo.

Memories of my first meeting with Buluman — his intelligent, brown eyes and that unique, pungent gorilla smell — are very firmly etched in my mind. Buluman, like all gorilla silverbacks, looked positively awesome but was the proverbial gentle giant. In a gorilla colony the silverback is in total charge. He is a benign dictator, the protector, the peacemaker, the lookout, the leader, the lover and husband and the patriarchal disciplinarian. With only old Betsy to lead and love at Taronga I don't really think Buluman's fine qualities were truly utilised to the fullest. Betsy appeared to me to be somewhat of a screamer who seemed to spend her day nervously whingeing about every little thing. I do believe, however, that Buluman's patience and his diplomacy were often stretched to the limit by

Betsy's somewhat neurotic behaviour, but he always seemed to manage to maintain his outward calm.

The Zoo's Head Keeper, Graham Button, had the rather enviable task of showing new arrivals like me around the fascinating behind-the-scenes areas of Taronga in their first week. Graham's incredible knowledge, gathered over many years of zoo keeping, was a godsend. His pride in Taronga and his important role there was very obvious to me, right from that first tour with him.

One of my privileges on this tour with Graham was to feed an early morning jug of warm milk to Buluman the gorilla. It was my first close encounter with a zoo animal and I recall being absolutely in awe of this huge creature. I was also secretly very scared, but there was no crossing the road to avoid this one! I just smiled and looked like I was enjoying it. It didn't take long for me to realise that in this job, I would be confronted with many situations that would test my courage and my nerve. But I soon discovered it would also give me so much in return. Animals teach you great patience and true zoo people seem to have this unique quality. That first morning it dawned on me what an extraordinarily privileged position I was now in. How many people could start their working day by saying good morning to a gorilla? That sense of privilege never left me, even after 25 years.

Despite being extremely nervous at meeting my first-ever gorilla — and at such close quarters, too — I was very impressed with Buluman's exceptional manners, and he soon put me at ease. He appeared to me to be such a gentleman. His enormous hands took the plastic jug from me so carefully and he drank the warm milk ever so slowly, savouring every drop. He then handed the jug back through the bars for a second round — no slurping and no spills.

BULUMAN THE GORILLA FEATURED A few months later in a rather obscure 'scientific' experiment which Zoo veterinarian Ted Finnie declared (I think rather tongue-in-cheek) would determine the animal's favourite colour. I hasten to mention that the results of this experiment never appeared in any scientific papers.

The Sydney Hilton Hotel sponsored Buluman and, as his birthday was approaching, we were asked to consider what kind of birthday gift the hotel could present to the gorilla. Taronga's Director from 1976 to 1979, Dr Peter Crowcroft, had brought the zoo animal sponsorship concept with him from Brookfield Zoo in Chicago, and introduced it to Sydney's Zoo in 1977. This new idea, which encouraged the community to become much more involved with the Zoo and its animals, quickly became very popular with all kinds of people.

Soon Taronga had an extended family of sponsors and supporters who had a personal passion for animals. Some people sponsored animals because they looked like someone they knew. Others who lived in apartments and who could not have pets joined the sponsorship program and visited regularly to see their Zoo pets. Company, club and community groups whose logos featured certain animals all joined in. From these humble beginnings there was a huge groundswell of support for Taronga and subsequently Western Plains Zoo. Over the years, this innovative fundraising program has gone from strength to strength.

We all put on our thinking caps and decided that the only gift a gorilla could safely play with was a large and indestructible tractor tyre. The Hilton's public relations manager Sue Gallie thought it should be painted up brightly in Buluman's favourite colour. But how were we to determine this?

This is when Ted Finnie sent me scurrying up to Mosman Junction shops to buy six different sheets of brightly coloured wrapping paper for part of an experiment. We locked Buluman in his night-den and placed the six different pieces of coloured paper on the floor of his display yard. On top of each piece of paper we placed a banana — a favoured food item of the gorilla. Ted declared that whichever banana Buluman went to first, the paper it was on would be the gorilla's favourite colour. I remember thinking how clever the vet was to develop such a theory.

We retreated to the observation area to watch Buluman emerge from his den. It was obvious that he knew something was up. He looked at all the goodies in his yard and, after careful consideration and thought, went up to the closest banana and gently took the paper from under it. He then went to the second banana and the third banana and did the same until he had collected all six pieces of coloured paper. Without further hesitation, he put all the pieces of coloured paper in his mouth at once and swallowed them in one gulp, completely ignoring the bananas, which remained forlornly on the ground!

So much for that experiment. Buluman was ultimately given a yellow tyre because we had a lot of yellow paint at the time. I should mention that this experiment was done in the pouring rain, and as we human apes stood on the outside of the bars getting drenched, the rainforest-dwelling gorilla was dry and warm. I can't help thinking that Buluman may have enjoyed a bit of a game that day at our expense.

He may have also just had a thing about the taste of paper because another time he demonstrated his liking for the wrapping and not the 'goodie' inside — or was Buluman just playing games with us again?

IT WAS JANUARY 1977 and Sydney was in the grip of a heat wave. This usually meant I would receive a call from one or more of the media outlets wanting a 'Zoo hot weather story', so I always made sure I had a few ideas up my sleeve.

Graham Button told me that dear Buluman loved ice creams, particularly vanilla Drumsticks with chocolate and nuts. I could see a 'hot weather photograph' in the making. This was totally inappropriate food for a gorilla, but agreed to by the Head Keeper and veterinarian on rare occasions.

The *Sydney Morning Herald* photographer Rick Stevens — very much a Zoo enthusiast — was always looking for that definitive animal shot. Rick quickly took up my offer of Buluman enjoying an ice cream as a perfect photograph to illustrate a hot Sydney day.

I bought six Drumsticks out of the meagre public relations entertainment budget and took Rick to the gorilla enclosure, where Graham let us into the service area between the outside wire and the inside bars of the exhibit. We handed the first Drumstick through the bars to Buluman who took it gratefully then walked back a few paces and swallowed it whole, paper and all, in one gulp. Rick didn't even have time to focus.

Buluman repeated this rather uncharacteristic, greedy behaviour with three more ice creams and by the time we were handing over the fourth, I was getting somewhat worried that Rick would not get the promised shot. Buluman was not demonstrating his usual impeccable manners and was clearly getting a little impatient with this 'game'.

While Rick was busy concentrating on what he could see in his frame, and I was obviously thinking about how much I would like an ice cream, too, Buluman suddenly put his huge hand out through the bars and grabbed the strap of Rick's spare camera, pulling camera and photographer flat up against the bars. Rick is not a large person and I had a scary vision of him being pulled through the bars into Buluman's domain. It should never be forgotten that Zoo animals are not tame; they are still wild creatures which perhaps have lost their fear of humans. Gorillas are incredibly strong and have large teeth which could inflict enormous damage. This incident gave us both an awful fright. Rick went pale while I tried not to panic. Buluman just looked annoyed that no more Drumsticks were forthcoming.

Luckily, I had one left and quickly handed it over to Buluman, who looked very pleased with himself. He released Rick and wandered off to enjoy this last ice cream. This time, he neatly unwound the wrapping and sat for 15 minutes, licking the ice cream, picking off the nuts and sucking his fingers which were sticky with the melting chocolate. He was very obviously savouring the flavour of this treat on such a hot day. Once poor Ricky stopped shaking he got a great shot, which appeared on the front page of the paper the next morning.

I am so very grateful that Rick did not come to any harm. Over the next 20 or so years, he became one of my favourite 'zoo animals'. Rick's regular visits produced hundreds of truly magical photographs at both Taronga and Western Plains Zoos. His quiet approach and exceptional patience to just wait and let the animals do their own thing always rewarded him with powerful images. This gave the Zoo immensely valuable coverage in the *Sydney Morning Herald*, and wonderful photographs for me to use in annual reports, Zoo brochures, posters, the souvenir book and for the vital interpretative graphics.

THE UNIQUE TARONGA LOCATION, and the many and varied Zoo animals were frequently used in television commercials.

It provided another opportunity for us to portray the beauty of animals to a huge audience and gave Taronga a higher profile, which in turn increased much-needed revenue. I was approached by a production company one time who wanted to use the gorillas and the gorilla exhibit to film a commercial about a new camera, which was meant to be 'so easy to operate even a gorilla could use it'.

It was much too dangerous for Buluman to be given a camera to play with as he would smash it in a second and could hurt himself on the sharp pieces. Instead, the production company had a gorilla suit made by special effects artist Bob McCarran. The idea was that the real Buluman would be filmed looking at an actor dressed as a zookeeper who was sweeping and cleaning outside the gorilla exhibit. The keeper would take the new camera from his pocket and photograph the gorilla. He would then hand over the camera to the gorilla — Bob wearing the gorilla suit — who would be seen taking a photograph of the zookeeper.

Bob is an exceptionally talented artist who is responsible for many lifelike animal characters we see in films and on television. In his preparations for the production of the gorilla costume, he came to Taronga many times to study Buluman's appearance, movements, character, behaviour and habits. I am sure Buluman was studying this regular visitor at the same time as well.

Bob's detailed research and amazing skill produced a costume of latex rubber and individually implanted yak hair, which bore an uncanny likeness to dear Buluman. There was even a remote-control device which enabled the facial muscles to move like a gorilla's — all that was missing was that unique body odour.

At 4.00 a.m. on the day of the filming, Bob used my office to change into this extremely realistic suit. After applying the finishing touches of brown contact lenses and smudgy dark eye make-up, it had taken a total of four hours to put on, and filming commenced at 8.00 a.m.

I forgot, however, to warn the Zoo's night watchman that there would be a 'gorilla' in my office in the early hours of that morning. On his dawn rounds, a poor security guard thought he had come upon his worst nightmare — a gorilla escapee who had set up home in the Public Relations office.

The television commercial was ultimately a huge success and lots of cameras were sold as a result, but I think that night watchman retired from his job at the Zoo not long afterwards.

As I have mentioned before, Buluman and his partner, Betsy gorilla, had been together at Taronga since the early 1960s but, despite being compatible, had never produced any offspring. In fact, veterinary tests later determined that dear old Buluman shot blanks.

In 1980, Zoo Director Jack Throp made the controversial but courageous decision to send Taronga's popular gorillas to Melbourne Zoo for the sake of the species. Jack Throp was Director of Taronga and Western Plains Zoo for seven years from 1979 and he was a kind, creative and forward-thinking American. There were far better housing facilities at Melbourne Zoo and as gorillas are social animals that live in large family groups, it would be much better for the Taronga pair to join the gorilla group in the zoo in Victoria. Buluman would never add to the gorilla population, but his very impressive presence might just inspire a younger male gorilla to exert himself.

Although they were greatly missed by Taronga's visitors and staff alike, I am very happy to say that Buluman and Betsy's new life suited them both exceptionally well. Whenever I visited

Melbourne Zoo during the ensuing years I always took time to stop by and see my old friend Buluman, who had settled into the Melbourne gorilla group and lived a comfortable life with them in the wonderful new gorilla habitat. He died in March 1998 aged about 40 years.

Betsy stopped her screaming and shouting and actually got herself a younger man. She subsequently produced two of Melbourne Zoo's much celebrated baby gorillas — Buzandi, a male, was born in December 1991, and Bambuti, a female, was born in April 1994. Judith Ainsworth Henke, Melbourne Zoo's Communications Manager and a long-time zoo 'sister' of mine, tells me that Betsy, although now almost 40 years of age, is still going strong and her only concession to her advanced years is that she is taking medication for age-related high blood pressure. She still plays an important role as auntie and sometimes surrogate mother for the younger gorillas at Melbourne Zoo.

As part of Melbourne Zoo's commitment to the international breeding program for Western Lowland Gorillas, Betsy's offspring Buzandi and Bambuti were sent in 2001 to join other gorilla groups in Hanover Zoo, Germany, and the Durrell Wildlife Conservation Trust on the Channel Island of Jersey.

THE TRANSFER OF BULUMAN and Betsy to Melbourne Zoo gave Taronga an opportunity to knock down yet more of the old concrete and wire of the 1940s and make space to build something more in step with the zoo of the 1980s.

However, just before the enclosure was finally demolished, a human ape moved in. Performance artist Mike Mullins lived in the old exhibit for one month as part of a Festival of Sydney event in 1984.

The human animal was a very popular Zoo exhibit during those January school holidays. We established set 'talking times', similar to animal feeding times, when Zoo visitors were able to converse with the display human and ask him deep and meaningful questions such as 'What do you eat?', 'Where do you go to the toilet?' and 'What do you do at night?' Taronga and Western Plains Zoos spend millions of dollars in their endeavours to display the world's most exotic wildlife, but here was a perfect example of people being fascinated more by their own species. Huge holiday crowds flocked to see this strange new animal, sending the Taronga turnstiles spinning and giving us a very healthy bottom line at the end of the month.

Mike Mullins was a delight to have as part of the animal collection. He was a very well-behaved creature, kept his enclosure clean and tidy, was exceedingly patient with the many media requests and visitor questions, and great fun to have a drink with after work.

He didn't come out of his enclosure for the whole four weeks, so he really enjoyed the company of Zoo staff or his friends for a drink or a coffee when they finished their working day. Mike confessed that he was a little bit scared after dark, as the nocturnal Zoo sounds were very unfamiliar to him. Many a night he was woken by possums scurrying across his 'bedroom' floor or by peacocks screeching at dawn, and Mary the gibbon's early morning hoots often startled him. But I think it was the lions roaring in the pitch dark that spooked our human ape the most.

The Zoo was sorry to see Mike leave but I think he was very grateful to be eventually rehabilitated and released instead of perhaps being sent to an international zoo on a breeding program!

GORILLAS HAVE A VERY strong attraction as a zoo exhibit and visitor surveys over the years indicated that Sydney really missed these handsome apes at Taronga. As a result, the mission to find a new gorilla group began in earnest in the early 1990s.

Dr John Kelly, a veterinarian by training and then a career public servant, took over the leadership of both Zoos from Jack Throp in October 1987. John initiated an exhaustive search a few years later, as only the tireless John could, for a suitable group that would be available to come to live at Taronga as part of the World Zoo Conservation Strategy for Western Lowland Gorillas.

A huge effort by all Zoo staff — especially Senior Curator William Meikle and Deputy Director Glenn Smith — saw the realisation some five years later of John Kelly's plan to again have gorillas as part of the Taronga collection. A new Gorilla Forest was built with state-of-the-art facilities. In December 1996 a Zoo team led by Glenn Smith, which included veterinarian Larry Vogelnest and African Division Manager Libby Kartzoff, travelled to Apenheul Primate Park in The Netherlands to safely escort the newly acquired and very precious family of ten gorillas to their new home at Taronga. Two keepers from Apenheul who the animals were very comfortable with also travelled out with the Sydney team to ensure the gorillas settled as quickly as possible. It was a well-planned and successful operation for possibly the largest-ever transfer of gorillas.

Kibabu, the silverback and head of the new gorilla family, was born in England in 1977 and then moved to Holland. His magnificent family now comprised of adult females Moila, Frala and Kriba, juvenile males Haoko and Shabani, juvenile females Kijivu, Shinda and Anguka and tiny baby girl, Safiri, who was only six months old.

They arrived at Sydney Airport on a pouring wet Saturday evening, but the welcome they received was warm and

Gorilla Shopping List

TARONGA'S WESTERN Lowland Gorillas have a wide and varied diet. Every week the gorilla family's shopping list from the vegetable market alone includes:

1 KG APPLES	14 PACKETS MUNG BEANS
2.1 KG BANANAS	500 G ONIONS
7 KG BROCCOLI	1 KG ORANGES
14 BUNCHES BOK CHOY	1 KG PARSNIPS
1 CABBAGE	4 BUNCHES PARSLEY
70 CAPSICUMS	1 KG PEARS
14 CAULIFLOWERS	1.6 KG PEAS
70 BUNCHES CELERY	1 PINEAPPLE
28 BUNCHES CHICORY	1 PUMPKIN
5 CORN COBS	14 BUNCHES SHALLOTS
28 CUCUMBERS	28 BUNCHES SPINACH
14 BUNCHES ENDIVE	2 PUNNETS STRAWBERRIES
14 BUNCHES LEEKS	7 KG TOMATOES
1 KG KIWI FRUIT	1 KG TURNIPS
140 KG LETTUCE	1 WATERMELON

In summer, Kibabu and his family of Western Lowland Gorillas at Taronga Zoo receive frozen watermelons and mango iceblocks as special treats. The gorillas also receive large quantities of fresh browse (tree branches and leaves) daily, some dried fruits, nuts and seeds and vitamin C supplements.

enthusiastic and there was not a dry eye among the Zoo staff who were waiting there to greet the travellers. We all said it was the rain on our faces, but I know mine were tears of both joy and relief that these gentle apes from across the other side of the world had touched down safely.

Zoo staff always have a huge responsibility for the animals in their care — especially at times like these — and I could see the strain of the journey on the faces of Glenn, Larry and Libby as they carefully supervised the unloading of the animal crates. I am sure the gorillas had much more sleep on that long flight than the accompanying staff did.

Channel 10's Lyndal Davies produced a comprehensive one-hour documentary titled *Gorilla gorilla*, which graphically told the story of the building of the new Gorilla Forest, the training of Zoo staff, the gorillas long journey from Europe to Australia and the amazing amount of detail, care and personal commitment that goes into a Zoo project such as this.

I initially thought that having a television crew following your every move during such a delicate and sometimes tense transaction would be the last thing Zoo staff needed. However, the crew were very sensitive and great to work with, and Zoo staff realised the enormous value of such coverage as it gave local, national and international audiences an insight to behind-the-scenes at the Zoo.

The Zoo staff are multi-talented people and the media focus on the Zoos' activities requires many of them to be extremely savvy communicators. Effective communication via the media is a vital skill for Zoo staff, which can never be taken for granted. Wonderful friends in the media like Bruce Barnett, Bill Edmonds, John Raedler and Richard Morecroft over the years assisted us with media skills training for everyone, from our elephant keepers to our insect breeders and engineers. Regardless of this essential professional training, it is the passion for their subject and their commitment to the Zoo cause which always shines through.

A police escort brought the gorillas safely to Taronga's gates, and Kibabu and his family were unloaded into the night quarters of their new harbourside home.

This is when I got my first smell again of gorillas in Taronga. Even after 15 years, that familiar nutty, pungent aroma took me back to my early Zoo days and I fondly thought of Buluman and Betsy again and how much they would have enjoyed being part of this wonderful new group of gorillas at Taronga. We kept them under wraps for four weeks while they became acclimatised and completed their regulation quarantine period. The Zoo's public relations team had worked hard on the media strategy for the debut of the gorillas to coincide with the opening of the new Gorilla Forest.

The opening day for the Gorilla Forest on 3 January 1997 dawned hot and steamy. By the time the 200 invited guests gathered at Taronga for an African-style breakfast to celebrate this significant and historic event, the temperature and humidity replicated that of a central African rainforest.

Opening days are always fabulous but frantic, and this one was no exception. Taronga was buzzing with a noisy and colourful African drum band, demanding Environment Ministers, eloquent Zoo Board members, proud sponsors, excited children, jostling media, enthusiastic guests, demented public relations staff, emotional keepers and shy gorillas — the usual ingredients for a successful 'official opening'. It all went smoothly, and according to our well-rehearsed running order.

At one time I was running, two-way radio in hand, on an urgent mission to find the Minister or Zoo Director for a television interview, when I accidentally 'goosed' the major sponsor with the aerial of my radio. Being small, I was able to hide in the crowd and remain anonymous but would now like to apologise to McDonald's Managing Director, Charlie Bell, for this considerable indiscretion. It may amuse Charlie to know that I subsequently won the Zoo staff 'Galah' award for this 'rear guard assault' and was the laughing stock of that year's Zoo Silver Shovel Awards for Excellence. From that day on I always carried my two-way radio safely in a basket, along with my mobile phone, news releases and pen. I looked pretty silly skipping through the Zoo like some kind of elderly Red Riding Hood but I decided it would be very difficult to 'goose' anyone with a basket.

The births and subsequent deaths of babies in the Kibabu gorilla family in 1998 were incredibly high and low points of life at Taronga. The amazing media coverage of the births paralleled our joy at seeing two more Western Lowland Gorillas come into the world. Their tragic deaths a short time later had enormous impact across the Zoo, and the media was voracious in their appetite to cover these losses, keeping me and many other Zoo staff on our toes for days and nights on end. The public sympathy — the cards, flowers, phone calls and donations — were an indication to all of us of the deep regard the community has for Taronga, and of the incredible magnetism of the magnificent gorilla.

I was sad that I retired from the Zoos before I had an opportunity to see first-hand a Taronga-bred gorilla baby develop safely in our care there. I am certain, however, it will happen before too long and I wish Kibabu, his family and the zookeepers who look after them much success in the future.

* * *

I CAN'T FINISH MY reminiscences of the delightful gorillas on a sad note, so would like to relate my gorillas' 'chrysanthemum reject' story.

A couple of years ago, we decided we could achieve some positive media coverage for the upcoming Mother's Day weekend and for our beautiful gorillas, by getting the young gorillas to give bunches of chrysanthemums to the adult females Frala, Mouila and Kriba. Maria Finnigan and her team of African Division keepers all agreed that the gorillas would be very interested in the colourful flowers.

On the keepers' instructions, I located at the flower market chrysanthemums that had not been sprayed with any chemicals. All rubber bands had to be removed for safety reasons, so we tied the strong-smelling flowers with edible shallots instead. The media was invited for 11.00 a.m. and a good turnout arrived with eager anticipation. It was Friday, and such photogenic Zoo news always provided a warm and fuzzy finale to the evening television bulletins, and a good lead into the weekend for Saturday's newspapers.

The gorillas were put inside their night house, the media lined up on the walkway, the keepers spread the usual food and the large bunches of chrysanthemums around the yard and when all was in readiness, out came Kibabu and his troop. The gorillas immediately spotted the unfamiliar flowers and headed straight towards them, looking at them cautiously. Unfortunately, they instantly lost interest and concentrated on enjoying their more familiar fruit and vegies.

When all this had been devoured a few of the young gorillas sauntered over to the flowers and began poking them with their fingers. They sniffed at them but quickly backed off. None of them had the slightest inclination to pick them up, let alone 'present' them to their mother. I had fleeting but disturbing recollections of that long-ago gorilla and the ice cream story, and after about half an hour was starting to feel a little nervous that the promised pictures would not eventuate. I could see Rick Stevens hoping as usual to get one of his superlative shots for the *Herald,* giving me a very knowing look.

After finishing their morning feed, the entire gorilla family wandered off for a siesta as they would in the wild. They left their yard littered with very forlorn-looking bunches of yellow and white chrysanthemums which were beginning to wilt in the midday heat.

Suddenly Kijivu, one of the young female gorillas, ran over to a bunch of flowers, picked it up, clutched the chrysanthemums to her body, held them at arm's length, swung them over her head, showed them to another nearby gorilla, sniffed them, picked up several other bunches and did absolutely everything I had hoped for. It was as if she had suddenly remembered her lines. The cameras clicked and whirred and everyone went away happy.

GORILLA FAMILY UPDATE

YOUNG FEMALE GORILLAS SHINDA, aged 11, and Kijivu, aged nine, now live in Prague Zoo in the Czech Republic. They were transferred there in 2001 as part of the World Zoo Conservation Strategy for Western Lowland Gorillas. It is time for them to start thinking about a family of their own. I just hope that Kijivu isn't given a bunch of chrysanthemums at the birth of her first baby!

As we left the walkway I noticed out of the corner of my eye that Kijivu was now plucking off the flowers and chewing them. A look of utter disgust came across her gorgeous face as she registered the terrible bitter taste and she quickly spat the flowers into her hand. It was then that I decided that chrysanthemums are a quite ordinary flower and gorillas definitely deserve roses.

IF GORILLAS DESERVE ROSES, then the Zoo's chimpanzees, I am sure, would be demanding bananas.

Initially, I had no great interest in the Chimpanzees. My early publicity work for Taronga rarely took me to their part of the Zoo. Their housing — old 1940s-style cages with wire and bars — made these animals very difficult to see or appreciate. Because of this there was no media interest either, as they were too difficult to photograph, being dark animals in a dark environment. To me they seemed to be noisy, smelly, very naughty and somewhat scary primates that spent a lot of time pointing their rather large and often bright pink backsides towards the Zoo visitors.

Although the Zoo's records are a little incomplete around this time, Taronga has probably had a history of displaying chimpanzees since about 1920. I can recall seeing chimpanzee tea parties at the Zoo when I was a child in the late 1940s. These were the days when our hairy relatives were dressed up in suits, ties and hats or lace frocks and bonnets and were trained to perform human activities for our entertainment.

As Zoo philosophy regarding the display of animals gradually changed, such performances ceased. The chimpanzee group at Taronga was added to and also bred exceptionally well. It grew into one of the largest colonies of zoo Chimpanzees in the world, eventually outgrowing the old-fashioned chimp accommodation.

With the establishment of Western Plains Zoo at Dubbo in 1977 as an open-range sister zoo to Taronga, many of the larger animals housed in rather cramped quarters at the Sydney zoo were moved to this exciting new property in central western New South Wales. The ungulates or hoofed herd stock such as deer, antelope and bison travelled across the mountains to a new home out west, and the combined areas which these animals vacated at Taronga became a new home for the chimpanzees.

Chimpanzee Park, which opened at Taronga in August 1980, is a huge grassed paddock about the size of a large football field. It's landscaped with a stream, waterfall, rocks, a fallen forest for climbing and shelter and large palm trees for shade — a quantum leap from the concrete and bars of the zoo of yesteryear.

Sixteen of the 21 chimps were born in Taronga's old exhibit, so, for the first time in their lives they would feel grass under their feet, see the sky directly above them, get 'buzzed' by birds, enjoy chasing turtles that lived in the moat, and see Zoo visitors from a totally different perspective across water instead of through bars. An air-conditioned night house, with shelves and ledges perfect for nest building, would keep them safe and warm while they were sleeping.

Channel 10 filmed a documentary *Taronga's Chimps Take to the Trees*, which was presented by John Laws. It graphically detailed the construction of the new Chimpanzee Park, the chimps' former life and their move from their old exhibit to their new home. Over the four weeks of this exacting chimpanzee transfer process many funny and emotional events took place, which were all recorded in this documentary.

Initially all the chimpanzees were anaesthetised with a dart and moved either individually or in family groups to the veterinary hospital where they had a full medical checkup. After this, the animals were moved systematically into their new night house, the more submissive chimps going first. To everyone's delight just a week after moving in, Spitter, one of the older female chimps, gave birth to a baby daughter. Once the chimps were confident and at ease in their new night quarters, they were slowly released into the paddock display area. Again, the more submissive animals went first, allowing them to explore and establish themselves in relative peace before the more bossy chimps — especially the alpha male — arrived. Staff gathered on the walkway early one morning before the Zoo opened to witness the first outdoors adventures of these chimps. There were many tears shed as several adult female chimps and their clinging offspring nervously ventured out into the wide, open spaces of their new home.

As the name *Taronga's Chimps Take to the Trees* implies, the Chimpanzees needed to be happily swinging in the trees at the conclusion of the hour-long documentary. The film crew had waited about two weeks for the reluctant chimps to perform, and by now there were only two days until the producer, Bill Edmonds, needed to have the footage to the edit suite to meet his on-air deadline.

The nervous chimps were very hesitant to explore their new multi-million-dollar home and no amount of encouragement by keepers would see them take to the trees. To say that Bill was a little nervous was an understatement.

As I so often observed, animals teach you great patience. Primate Keeper Brian Brettle offered to try one last time to fulfil the promotional promise of the documentary and ensure his precious chimps could be seen happily swinging about in their new trees. Not only did we want to deliver what we were about

CHIMPANZEE FACTS

THE CHIMPANZEE (*Pan troglodytes*), like the gorilla, is the closest living relative to humans. Like humans, chimpanzees are social animals. They are only found in the tropical forests, woodlands and savannahs of west and central Africa, where they live in large, politically complex groups of up to 50 animals of all ages, ranging from newborn infants to adults of 40 to 45 years. It is the dominant or alpha male's role to protect his colony of chimpanzees and keep this often-unruly group in order. He is forever fighting off approaches from other would-be rulers and challengers. Communication within the group of chimpanzees is achieved through loud and varied vocalisation and through gestures, postures and numerous facial expressions.

Chimpanzees are very good climbers due to their powerful arms and specially designed feet and hands, which are perfect for grasping and swinging in trees. They are omnivores, which means they eat both plant material and some meat. They mainly feed on leaves, seeds, bark, insects and birds' eggs and are known to hunt small mammals using teamwork and tactics to catch this prey.

Like the gorilla, the chimpanzee is also now considered an endangered species. Their numbers in the wild have diminished because of forest clearing for

> agriculture, mining and logging and because they are hunted for the pet, bush-meat and circus trade. They have also been captured for use as laboratory animals and have felt the terrible effects of the ravages of war.
>
> Like the human ape, the chimpanzee can breed at any time, with the pregnancy lasting nine months. Infant chimpanzees are dependent on their mothers until they are four years old. During this time they learn the many things needed to be a happy and healthy chimp and to be an accepted member of the colony. They also learn tool-use to fish for termites and to crack open nuts.

to advertise, we all wanted desperately for the chimps to discover and enjoy the wonderful new world we had just given them.

Just before the chimps were released into the exhibit again, Brian climbed up every tree and put bribes of honey, bananas and strawberry jam along the branches and quickly retreated. The chimps all sat along the back wall as they had done for weeks now, clinging pathetically to the cold bricks, which were securely reminiscent of their old enclosure. But this time we held our collective breath as we noticed that a few of the more adventurous animals were curiously eyeing the treats in the branches; then they finally began to climb the trees.

Right at the critical moment when several chimps were getting into full swing, to our horror as much as the chimps', a royal 21-gun salute blasted across the harbour from Mrs Macquarie's Chair!

The chimps took immediate fright and ran back to the safety of their night house, refusing to come out for the rest of the day. We had to repeat the whole procedure the next morning, this time successfully, minus the royal cannon fire.

The chimpanzees became instant stars after the screening of Bill Edmonds's documentary, which coincided with the official opening of Chimpanzee Park in the August school holidays. The time I spent with them during the filming of the documentary, and the subsequent media attention, enabled me to develop some sort of understanding of these complex apes. I gained a huge respect and appreciation for their politics, and their close-knit family lifestyle, and became firm friends with the individual animals such as Lulu, Lucy, Susie, Fifi, JoJo, Sutu, Spitter and the entire chimp gang. We subsequently filmed a second documentary about Taronga's chimps, this time produced and presented by John Collis for Channel 9. This gave me even more reason to spend time with these amazing animals. It is easy to understand why people like British conservationist Jane Goodall devote a lifetime to studying and helping these animals who are in such desperate plight in the jungles of Africa.

OVER THE YEARS, I became particularly fond of Lulu the chimpanzee, who is quite a character. She had come to Taronga from a circus in the United States in 1965 when she was about 13 years old. A short, stocky chimp, Lulu enjoys a high ranking in the group. There are so many stories I could tell about this special chimp, like the time she was in love with our vet or how she has the amazing ability to take an anaesthesia dart apart and put it back together again before she loses consciousness. Once she used a forgotten scrubbing brush to clean her own night quarters and on another occasion watched 'Daktari' on television while recovering from abdominal surgery. A whole book could be devoted to the life and times of Taronga's chimpanzees, but I will restrict my chimp reminiscences to one special story about my friend Lulu that I absolutely love.

Lulu has always been fascinated by fireworks displays. Even in her old enclosure, which was closer to the harbour, she would excitedly watch the New Year's Eve performance through the windows at the back of her night-den. From Chimpanzee Park there is an even better view and Lulu can watch the boats on the harbour, too. She soon worked out that lots of boats gathering meant it was fireworks time again.

For days leading up to New Year's Eve 1999, thousands of boats congregated. Lulu had never seen a sight like this before and she knew she was in for a great fireworks display. In fact, by dusk that evening over 6,000 craft had packed the harbour. They contained more than 100,000 people also intent on seeing the much-lauded fireworks spectacular.

The slopes of Taronga were a popular vantage point to see in the New Year, and the African Division keepers wanted all chimpanzees safely tucked into their night house by nightfall, but Lulu had other ideas. She gathered her little family around her and they all stayed outside in the display yard, refusing to budge, not even being tempted inside by the special treats and food bribes their keepers were offering.

Now, old Lulu had a bottom swelling at that time which meant she was in season, so she was very attractive to the opposite sex. Poor inexperienced Gombe, a young male just beginning to sow his wild oats, decided that Lulu looked extremely attractive that night, so he decided to stay out in the display yard with her and her family, hoping for some other sort of fireworks for himself.

> ## Chimpanzee Security
>
> **CHIMPANZEES ARE DANGEROUS ANIMALS** and are treated with great respect and caution at Taronga. The humble banana is used as an effective security measure in Taronga's chimpanzee night house. Chimpanzees have large appetites and a male chimp can eat up to 50 bananas in one meal. Each evening before leaving for home, the keepers lock these animals inside for the night and put a banana on the inside of an observation window which looks from the keepers' area into the chimps' night quarters. In the morning, if that banana is missing, the keepers know immediately that a chimpanzee has somehow found its way out of the night quarters and into the locked corridor, as the first thing a chimp would do is eat this very appetising and accessible banana.

When the amazing harbour fireworks commenced with the loudest of bangs and bright explosions, which lit up the sky, poor Gombe got the fright of frights and ran shrieking into the security of the night house, completely forgetting his intentions. Lulu, however, lay back on the grass, soaking up the amazing spectacle and enjoyed her best New Year's Eve ever.

In recent years, the number of chimpanzees in the wild has declined drastically and they have vanished from nine of the 24 countries in Africa where they once lived. Former Taronga Zoo primate keeper Debby Cox is now the project director of the Ngamba Island Chimpanzee Sanctuary and Wildlife Conservation Centre in Entebbe, Uganda. Debby and the equally tireless Dedee Woodside have lured many talented and dedicated Taronga staff and Zoo Friends Volunteers to the island to lend their special skills to help make a difference for chimps. This sanctuary for orphaned, confiscated or relocated chimpanzees, which was established by a group of six international wildlife conservation organisations

CHIMPANZEE SPECIAL MENTION

IN 1982, SUSIE THE CHIMPANZEE, a resident of Taronga Zoo since 1953, was diagnosed diabetic. She was 34 years old at the time and from then on needed a daily shot of insulin. At first she was less than co-operative at injection time, but very soon learned that a banana reward was hers if she was a good chimp. Every morning when the vet or vet nurse arrived she very obediently presented her back to the bars of her sleeping quarters, patiently received her insulin shot and quickly put her hand out for the banana. Suzie's condition was monitored for 15 years in an entirely honorary capacity by Dr Shailendra Sinha, a specialist in human diabetes who took a genuinely keen interest in this case, and often described Suzie as one of his favourite patients.

(including the Jane Goodall Institute), is just the beginning of a broader charter to save the chimps of Uganda. It is very heartening to see that the skills, knowledge and expertise gained from observing and caring for zoo chimps can be used so effectively to help their wild cousins whose futures are not nearly as secure.

THE STORY OF TARONGA'S chimpanzees moving from their old, cramped quarters to a spacious new home was a prelude to a similar adventure, some 14 years later, for Taronga's orang-utans, who went from 'bars to branches' in March 1994 when the new Orang-utan Rainforest opened.

It was personally very rewarding to see gentle giant Archie the orang-utan swinging around his new rainforest home. The new exhibit provided the orang-utans with about five different options in their day. They had heated night-dens to accommodate their intricate evening nest building, an indoor exercise yard, an indoor display yard with lots of toys for use during wet weather and, of course, the wonderful outdoor rainforest. All of this was surrounded by lush rainforest trees and plants filled with real and recorded animal sounds, mist sprays and creative interpretive graphics depicting all facets of rainforest life, highlighting the urgent need to conserve these precious ecosystems.

I HAD ENJOYED MANY cuddles from Archie and his half-brother Moe when they were newborn orangs at the time I began working at Taronga in 1975. Archie had been born at Taronga a few days before I arrived and little Moe a couple of months before that.

During the late 1970s, Taronga Zoo had a fortnightly segment on a Channel 7 children's program called the 'Funshine Show'. The existing — or maybe they were non-existing — occupational health, safety and quarantine regulations enabled us to take a variety of Zoo animals to the studio at Epping so zookeeper Brian Brettle could show them and talk about them with the presenter. Brian and I would load up a Zoo minivan with a different mammal, bird or reptile each time and drive 30 kilometres to the studio. Much stricter regulations today do not permit such excursions with Zoo animals, particularly the non-native species. This popular show was, however, an excellent opportunity to promote the Zoo and educate its young audience about animals.

Finally the long-running show came to an end, and I have to say that Brian and I were quite pleased as we were starting to run out of transportable creatures. But we had kept the best for last and for the grand finale I obtained approval to enable young orang-utans Archie and Moe to make a surprise appearance. In preparation for their journey, Brian gave the inseparable Archie and Moe some practise drives around the Zoo grounds, to get them accustomed to the travelling conditions. I also made arrangements with the producers for Brian and his orangs to be first on the show so there would be no waiting around for the young animals.

The car trip went smoothly and Archie and Moe occupied themselves by peering out of their crate, full of wonder at all the new sights. We no sooner arrived at the studio than we realised things were not going according to plan. There had been massive production delays, lighting and equipment malfunctions, power failures and non-appearance of several of the other guests. So there we were with our crate of orangs in a dark studio with nothing to do but wait. The ever-inquisitive Archie and Moe were getting extremely bored and so were Brian and I. We were about to say we would, for the animals' sake, need to depart when everything

Orang-utan Facts

ORANG-UTAN IS A MALAY WORD usually described as meaning 'man of the forest', although I have also seen it translated as 'person of the forest'. The orang-utan (Bornean Orang-utan *Pongo pygmaeus* and Sumatran Orang-utan *Pongo abelii*) originates in the remote forests of Borneo and northern Sumatra, where genetic research has now led to the two distinct species being recognised.

The orang lives predominantly in the trees, using its strong, grasping hands and long arms to travel effortlessly through the upper, middle and lower storeys of the forest, swinging from branch to branch. Sometimes the heavier orang-utans, especially the huge males, need to find a more secure way to move around and descend to the ground to move between trees.

The orang-utan is a natural gymnast. A male's long arms can span up to 2.2 m, enabling it to swing effortlessly from tree to tree. The orang-utan also has extremely flexible hip, wrist and shoulder joints, which give it a far greater range of movement than other great apes.

They are quite like gorillas in their eating habits, grasping fruit, leaves and vegetation with their hands and feet; however, they will occasionally eat eggs and insects, lizards and small mammals. To obtain water, they lick rainwater from plants and from their own

> long hair. They build nests in trees to sleep safely at night and are relatively solitary animals, the only groups being that of mother and infants.
>
> While the orang-utan is now a protected animal by law, infants are still captured and sold illegally as pets. The massive and ongoing destruction of their rainforest habitat by the human ape is the major threat to their survival in the wild.

started working and it was on with the show. Brian got the orange bundles out of the crate, only to see the baby orangs leap from his arms and climb up the nearest lamp stand and to the top of the stage set. With only seconds to spare before recording commenced, the agile Brian somehow retrieved his charges from their lofty perches and he and they made a memorable but rather breathless appearance on national television.

Moe moved on to Barcelona Zoo in September 1982; but Archie stayed at Taronga, where he grew into a very handsome adult orang. He had a long, raggy red coat trailing with amazing dreadlocks, huge cheek flanges, a fine beard and moustache and sensitive, inquisitive brown eyes. Archie deserved to be appreciated in a more natural habitat than his old concrete and barred accommodation could provide, and now looked very handsome hanging out in his new Zoo 'rainforest'.

MARGARET MILLER, THE ZOOS' amazing Archives Manager who understands the depth of zoo history pointed out to me that Taronga's records include brief mention of an orang-utan being part of the collection at the old Moore Park Zoo around 1909, and, have been displayed at the Mosman Taronga since 1920. While I am sure that all Archie's antecedents were as special as he was, I wonder if they were ever as spoiled, or whether they had his penchant for French pastries?

Zookeepers try not to become too attached to their particular animals but it must be very difficult at times. I think Archie, a hybrid Bornean/Sumatran Orang-utan, was everyone's favourite. When he became ill with a chronic chest problem in the winter of 1994, he was given expert care by renowned St Vincent's Hospital cardiologists Dr Phillip Spratt and Dr Anne Keogh. These two thoracic specialists were also treating Zoo Director Dr John Kelly, who always said that only once they had 'experimented' on him would he allow them to treat Archie the orang.

The orang-utan keepers went to extraordinary lengths to ensure Archie took his prescribed medicine, which followed exploratory surgery. Each day he had to take antibiotic syrup, which the keepers gave Archie in a spoonful of strawberry yoghurt to conceal its rather sickly taste.

I was at the orang house with a film crew one morning and saw the keepers having difficulty getting the medicine spoon back from Archie. His massive brown leathery hand was wrapped very tightly around the spoon and not even the offer of additional yoghurt would convince him to hand it over. Orangs are such clever, calculating animals that this spoon, if left in Archie's night-den, could have been used by him as a lever to prize open a lock or two, so you can see it was imperative that it was retrieved.

The orang-utan exercise yards are opposite the keepers' kitchen and food preparation area. A large observation window between the two areas enables the keepers to check on their animals without actually having to leave the kitchen. Suddenly it became very apparent to us why the usually well-mannered and accommodating Archie was being so stubborn and difficult that morning. He had spied, through the window, a delicious-looking fruit flan on the kitchen table and his mouth was watering. He had cleverly worked out, as only an orang-utan could, that if he hung onto the spoon long enough he may be offered some of the flan as a bribe. He had a bargaining tool and he was going to use it. The flan, bought at considerable expense from the local French patisserie, was intended to celebrate Head Primate Keeper Paul Davies's birthday. Instead, Archie enjoyed most of it before he eventually handed over the spoon.

Sadly, no amount of tender loving care could save the very ill Orang-utan and Archie died in March 1996.

THE CHANGES I SAW to the housing and display of the chimpanzees, the gorillas and the orang-utans are wonderful examples of the constant evolution that takes place at Taronga and Western Plains Zoos.

The need for change never diminishes and the opportunities are always there — this is part of the magic and magnetism of Zoo life. One of the biggest challenges was always the financial one. There was never enough money to make all the changes everyone wanted and expected. But Zoo people are resourceful creatures and all of the staff work extremely hard to secure the funding so that conditions for both the animals and the visitors are being constantly improved.

The Taronga Foundation is the latest fundraising drive and was launched on the day I retired. It was initiated by current CEO Guy Cooper to raise the ongoing funding needed to ensure that Taronga and Western Plains Zoos continue as world-class centres of excellence in environmental education and conservation. I trust it will enable them to do even more for wildlife, well into the 21st century.

I HAVE ALWAYS CONSIDERED MARY, a Mueller's Gibbon (*Hylobates muelleri*), to be queen of the Zoo, where she has resided since 1960.

Mary's early morning hoots, typical of the dawn chorus of calls between gibbons in the wild, herald the beginning of each new day at the Zoo. This old girl has had many adventures during her long and colourful life. For most of it she lived with her 'husband' Robinson, swinging about the Moreton Bay fig tree on the island in Gibbon Lake. She produced several babies with Robinson, but unfortunately none of them survived. She was not a good parent and mis-mothered her offspring. Motherhood just did not suit Mary, so eventually, in 1982, she had her tubes tied.

Sadly, Robinson passed away in 1986 and Mary was alone. In an effort to cheer her up, Taronga curators secured a new 'husband' for her later that year. Silver, the new male gibbon, was quite a deal older than Mary and came with impeccable credentials from Perth Zoo. Mary was not at all grateful for this matchmaking and gave poor Silver a terrible time. Veterinary records show that each animal was given 5 milligrams of Valium daily to calm them, as the 'getting to know you' period was extremely difficult and the relationship always stormy. Mary even swung onto low-hanging branches and with her arms outstretched and looking just like a tightrope walker, waded off her island and into the nearby gardens. She was somehow encouraged back to the marital tree but only to harangue and nag poor old Silver once more.

Gibbon Facts

THE GIBBON IS AN APE but is classified as a 'lesser' ape as opposed to the 'great' apes, which are the orang-utans, gorillas, chimpanzees and bonobos.

They are found in the forests of Southeast Asia and most species of gibbon are classified as endangered. They are small apes, with the adult males growing to about 59 cm in length and weighing about 7.5 kg.

This most acrobatic ape swings through the trees with an arm-hanging, hand-swinging method known as brachiation, which saves energy by maintaining momentum using the body as a pendulum. The Gibbon releases its hook-like grip with one hand at the height of its swing's arc. Its forward-facing eyes allow stereoscopic, distance-judging vision that helps to determine the next hand-hold, which may be up to 3 m away.

Not only is the gibbon an effective tree-swinger, the big toe on each foot enables it to grasp in opposition to the toes so that they can walk upright along branches. They are considered to make lifelong pair bonds, although more recent surveys have found that some changing of partners does occur. Sadly, gibbon habitat is being aggressively destroyed, and hunting by humans is also a major threat to their survival in the wild.

Finally, Silver couldn't stand it any longer, and took an almighty leap off the island and far away from cranky Mary. She was a 'swinging single' again and he spent the remainder of his days quietly in the Zoo's veterinary quarantine centre.

During a very severe winter storm in August 1990, Mary's Moreton Bay fig tree fell over with poor Mary still clinging to the branches. The huge 70-year-old tree fell across the lake and made a bridge for Mary to escape from her island. Instead of running off to explore other exciting parts of Taronga Zoo, Mary ran straight into the secure arms of her keeper Paul Davies, grateful to see a friendly face after such a terrible ordeal. I think it must be the first and last time that Mary has ever been pleasant to her keepers. She still mutters ungratefully at them every time they clean out her house.

The Zoo Friends generously funded a new but mature five-metre-tall tree for Mary and the gibbon is, once again, back on her 'throne' where she belongs. Her home is directly below Taronga's popular Seal Theatre and every day Mary can be seen swinging around her tree in time with the seal show's introductory music. I don't think she is too interested in those show-off seals, but she seems to thoroughly enjoy both the music and the milling crowds at performance time and, never one to be outdone, Mary puts on quite an acrobatic show of her own.

Mary's loud vocalising epitomises the variety of jungle sounds to be enjoyed every day at Taronga. Her haunting hoots can best be enjoyed early in the morning and as Radio 2GB was preparing to do an outdoor weekend live broadcast from the Zoo in October 1986 as part of Taronga's 70th birthday celebrations, I suggested that they choose a location adjacent to Gibbon Lake as their broadcast point. I guaranteed that the

MARY THE GIBBON'S SHOPPING LIST

IN RECENT YEARS, old Mary's appetite has decreased, but variety is the spice of life and her keepers make sure that every day she has a wide selection from which to choose.

Daily she is offered the following:

1 APPLE	1 CELERY STICK
2 BANANAS	1 SHALLOT
½ ORANGE	¼ CAPSICUM
½ PEAR	SMALL AMOUNTS OF
1 TOMATO	LETTUCE, CABBAGE
1 CARROT	AND PEAS
¼ SWEET POTATO	1 CORN COB WEEKLY AND A
¼ CUCUMBER	SMALL AMOUNT OF CHEESE
	AND PINEAPPLE

presenter, Bruce Barnett, could take advantage of Mary's unique 'good morning' calls — perfect background sounds for the radio. We all gathered at dawn but there was no sign of any life from Mary, so the presenter had to broadcast in total silence. We might as well have been in the morgue. The sun rose slowly over the Zoo, allowing us to see Mary in her tree, but we still could not hear anything remotely resembling a welcoming hoot from the cranky old gibbon. As I took a closer look I realised it wasn't her quaint old face she was showing us — it was her dark and hairy backside that she was pointing in our direction. We had obviously disturbed her beauty sleep and she was not amused.

When the program crossed back to the studio for the next news bulletin all the broadcast equipment and the presenter were bundled up and moved as quickly as possible in a Zoo truck to Discovery Farm where at least the pigs, roosters and 'Trigger' the donkey, sponsored by another radio man, Jonathan Coleman, provided a noisy welcome. Maybe Jono, on his frequent visits to see Trigger, gave the farm animals a few valuable tips about working in radio. Pity he didn't do some media training with Mary.

CHAPTER 2

Some of my Friends are Horny

O**N MY FIRST DAY** at the Zoo in May 1975 it was my honour to be involved in the media announcement of what was to be the last Black Rhinoceros calf to be born at Taronga. Even though I researched the background to the news release, I did not, at that time, have any idea of the significance of this Zoo birth, the perilous status of the Black Rhinoceros in the wild or of the future role of world zoos in conservation outcomes.

I can remember being amazed at this three-week-old grey baby which cantered around doing vertical take-offs and landing with a clunk on its still shaky legs. The events that made the most lasting impressions on me that day were the rhino mother's tender care for her baby, Head Mammal Keeper Dave Cody's pride in the birth (in fact, the calf was named 'Cody' in his honour), and the media interest in this latest Taronga birth.

RHINO FACTS

A BLACK RHINOCEROS (*Diceros bicornis*) adult can weight up to 2,500 kg and can measure around 3.5 m in body length. There are five species of rhinoceros. The Black and the White Rhinoceros comes from Africa, where hundreds of thousands of them used to roam. The other three species of rhino come from Asia. The rhino has been around for millions of years.

While they really differ little in actual colour, the Black Rhinoceros can be identified by its long, prehensile or pointed upper lip, which it uses to grasp leaves and branches while it is browsing for its food. The White Rhinoceros (*Diceros simus*) has a broad lip and muzzle specialised for grazing. The rhinoceros has excellent hearing and sense of smell but has relatively poor vision. Both the Black and White Rhinoceros have two hard, solid, fibrous horns in the middle of the snout, which are made of keratin, the same substance as hair, fingernails and claws. These grow about 7 cm a year.

Most Black Rhinoceros are solitary animals. A calf, born after a 15-month gestation period, will stay with its mother until it is about three years old, when the next calf is born.

Life expectancy for a rhino is from 30 to 50 years — that's if it is lucky enough in the wild to escape the rhino-horn poachers.

> The Black Rhinoceros has been part of Taronga's animal collection since 1938 when Sir Edward Hallstrom, then a generous animal enthusiast and later Taronga's Honorary Director until 1967, donated two such rhinos to the Sydney zoo.

I was captivated by this inaugural 'Zoo moment' and by the baby rhino, which looked like a grey bull terrier puppy born with feet and ears ten sizes too large. My involvement in that marvellous event on my very first day launched my subsequent and ongoing joy and wonderment of nature, which was nurtured on a daily basis at the Zoo.

Following my early infatuation with Cody the rhino calf, I subsequently came to know and respect Taronga's other Black Rhinoceros, too. There was the gentle male Ferdinand, who liked his back scratched with a key to the point where it made him go weak at the knees and he would buckle to the ground. Ferdinand was caught in the wild in West Africa in 1947. I also met the females: Beauty, born at Taronga in 1965 and subsequently the mother of Cody, and Taronga (named after the Zoo), born in 1958. I also got to know Dynah the rhino, an animal that had spent a solitary life in Perth Zoo until she was transferred across the Nullabor Plain to a new life at Taronga. Eventually in 1991 Taronga and Dynah were transferred to Western Plains Zoo. These two rhinos became the nucleus for the ambitious conservation program for Black Rhinoceros planned at that Zoo by the Zoological Parks Board of New South Wales.

Since that significant day for me in 1975, an estimated 65,000 Black Rhinoceros in Africa have been slaughtered for their horns, which are used as dagger handles and in traditional eastern medicines. An adult rhino's horn can weigh approximately six kilograms and each kilogram can fetch $6,000 to $10,000 on the black market.

There are now only about 2,400 of these ancient creatures remaining in the wild and a global captive population of 210. Numbers like these bring the Black Rhinoceros perilously close to extinction.

In 1989 the Zoological Parks Board of New South Wales, as a result of the vision of the Director and Chief Executive Dr John Kelly, joined the Zimbabwean Government and private wildlife conservancies in America to form the International Rhinoceros Foundation. Here was an opportunity for this State's two Zoos to look outside their walls and participate in a global conservation program. The aims of the International Rhinoceros Foundation were to assist the Zimbabwean Government with in-situ breeding programs and anti-poaching operations to try and arrest the illegal trade in rhino horn. The plan was also to set up viable Black Rhinoceros breeding programs out of harm's way in wildlife sanctuaries in Australia and the United States.

The world-class Black Rhinoceros Conservation Centre built at Western Plains Zoo, thanks to the generosity of the *Australian Women's Weekly*, is the largest of its kind in the southern hemisphere. The display area for the Black Rhinoceros provides an opportunity for visitors to learn about the plight of this animal and the Zoo's involvement in its future. The facility can accommodate up to 40 rhinos, and comprises extensive off-exhibit breeding and exercise yards complete with animal treatment and conditioning areas, shade huts, calving yards, mud wallows and high-pressure sprinklers. In a perfect world, the ultimate goal of this conservation effort would be to send Western Plains Zoo-bred rhinos for release into conservancies in Zimbabwe.

THE POWERFUL STORY of the capture of the Black Rhinos in Chete Safari Park in the winter of 1992, their quarantine on Cocos Island, the subsequent mammoth undertaking to transfer them to Western Plains Zoo and a little later the arrival of the four male rhinos from the United States was to be world news. I was extremely privileged to accompany the Australian media team to Africa for part of this dramatic story.

In June 1992 this media contingent, including a documentary crew from Channel 9 led by producer and presenter Tina Dalton, Reuters photographer Mark Baker, and the *Australian Women's Weekly* photographer Claver Carroll, joined Zoo Director John Kelly and myself on the Qantas flight to Harare, Zimbabwe, and subsequently on to Chete Safari Park on the shores of Lake Kariba. Five years earlier 250 Black Rhinoceros lived in this isolated, dry part of Africa. When we arrived only 15 rhinos remained — the others had been massacred for their horns. The rescue of these remaining animals was our mission.

I must say that I couldn't help feeling a little anxious about what might lie ahead of us on this rhino safari. I had never been to Africa before and although I was extremely excited about the prospect of visiting this ancient and fascinating country, the details of the trip did seem, at times, rather vague and therefore somewhat daunting. I think it was mainly because any map I had been able to acquire in Sydney before we set off made no mention whatsoever of our destination, this place called Chete. I really had no idea where I was going. I did, however, have great faith in John Kelly as the expedition leader and tried to put these niggling fears out of my head.

Our flight from Sydney, via Perth to Harare, Zimbabwe, was one of great anticipation for me as I contemplated what lay ahead. I also went over the long list of radio interviews John Kelly and I would do from Zimbabwe, which I had lined up with metropolitan, regional and national stations leading up to our departure. It was a strong conservation story and there was very encouraging media interest in the project.

After several days in the capital city liaising with the National Parks Service and other government authorities, completing a million forms and re-organising filming permits even though we had organised them all before we left Sydney, we were ready to depart from Harare by light plane for Chete. Film crews never travel lightly and we were no different, requiring two six-seater light planes, which suddenly became heavy planes, to transport all the gear plus the seven of us to our destination.

We gathered on the windy tarmac near our chartered aircraft and as the two pilots walked towards us, I immediately showed my age by commenting that one of them looked extremely young. I also declared that I hoped the older pilot flew my plane as he looked much more experienced. Mark Baker worried me by saying he would much rather have the younger pilot as 'he had much more to lose'. I decided that seemed a fairly sensible comment so stuck close to Mark as we clambered on board the plane with the 'Baby Biggles' at the controls.

It was an exhilarating flight and it was marvellous to have the second plane in sight the whole way. We were able to signal each other via the radios when we saw something wonderful on the ground below, such as a herd of elephants making a dusty red trail through the bush. Talk about 'Out of Africa': I felt, at that very moment, just like Karen Blixen!

After about two and a half hours flying time, we finally spotted a tiny, isolated, unsealed airstrip in the distance, surrounded by dense scrub and quite close to a huge body of water that turned out to be the fabulous Lake Kariba.

The two pilots began a decidedly intense discussion via the radio on the best way to land the planes on that tiny strip, and it was then that I wished I had gone with my initial intuition and flown with the more experienced pilot. It seemed our young pilot had never landed on that particular airstrip before and had to be 'talked down' by the older pilot, who said he would go first and show us how it was done. As we circled rather nervously, the first plane made its approach but completely missed the strip and had to pull out at the last minute and go around again. And this was the pilot who had landed there before.

Our young pilot, who happened to have Mark from Reuters sitting next to him as his 'co-pilot', began his approach, all the while receiving instructions from the one who missed the target. Needless to say, I had my eyes closed tightly for the landing, all the while wondering how I had got myself into such a situation. We made it down safely; our pilot looked very relieved and the 'co-pilot' looked a strange shade of green as he headed quickly to the on-board Esky for a beer.

Next, we loaded our mountain of baggage and camera gear onto two waiting National Parks Service trucks and clambered on board for a bumpy ride over dusty dirt roads to our base camp on the edge of beautiful Lake Kariba. This is when I got my first sighting of a herd of zebra wandering along a track and had the 'zoo thought': 'Gosh, they have escaped.' Then I spotted piles of elephant droppings and the zoo in me couldn't help thinking that the keepers here were a bit slack as it was way past 9.00 a.m. and it should all have been cleaned up by now!

Our camp for the next two weeks was a hunting lodge run by the National Parks Service for wealthy white tourists who had the strange urge to shoot things that moved. This program, however, did cull excess animals in the park, mainly antelope, and provided the Service with much-needed funds to assist with their conservation efforts.

The camp, despite being so remote, was quite comfortable and consisted of a line of thatched rondavels or sleeping quarters and a main house with kitchen, mess hall and bar. John Kelly, ever the gentleman, helped unpack all the gear from the trucks only to realise that everyone ahead of him had grabbed the rondavels closest to the main house. The remaining pathetic little hut, now his, was way out on a peninsula and looked very lonely and somewhat vulnerable. It suddenly looked downright dangerous when one of the local National Parks Service scouts, responsible to the ranger for the management of the safari park, pointed out that the rhino poachers, who came across Lake Kariba from Zambia at night, sometimes came ashore right near John's hut. They told us of strange footprints and marks left by their boat in the sand and the fact that the nearby generator had been stolen that week. I got that uneasy, nervous feeling again and lamely gave John my Swiss army knife for protection against nocturnal marauders. John, the bravest of men who was facing his own personal survival challenge at the same time as the rhino, gratefully took my knife and pretended it would make all the difference.

To my added dismay I soon discovered that our rondavels only had thin, flapping curtains — no glass — at the windows and a door that didn't lock, so none of us was going to be too safe from poachers or any other creatures.

The first night as I lay awake on my little camp bed I tried to concentrate on the events of the day while watching the huge African moon shine through my open window. I was, however, becoming increasingly distracted with the thoughts that a big cat of some description would jump through the window and maul me at any minute. Around 3.00 a.m., I began hearing deep grunts and growls and recognised them immediately as the sound of a leopard on the prowl — and it was very close, too. My mosquito net was the only thing between me and the leopard, which was surely lurking just outside my rondavel. I got up and put my suitcase against the door to at least 'leopard-proof' it, and I must have nodded off to sleep soon after because the next thing I knew the early morning wake-up drum sounded, alerting me to daybreak, and I leapt out of bed.

Around the campfire at breakfast the local scouts and rangers were describing how the hippos like to play in the lake at night, how they wander around the camp grunting and growling with pleasure as they feed and how they had been having a great time the night before doing just that. I didn't dare mention my prowling leopard theory.

THE FOLLOWING TWO weeks were filled with excitement and a good deal of tension as we trailed the National Parks Service Veterinarian Dr Mike Kock and National Parks Ranger Norman English through the scrub as they tracked and captured the Black Rhinos and prepared them for travel to their new life out of harm's way in Dubbo, New South Wales, Australia. I learned first-hand the harshness of life in Africa for both its wildlife and its people, and came to appreciate even more the vital role of the modern zoo in the conservation of the world's endangered species — so much more needs to be done before it's too late.

A bush trek on foot, early in our stay, revealed the pathetic carcass of a recently killed rhino. The dozens of skulls, hideous reminders collected and displayed at the scouts' camp, were also testimony that the poachers were very active and the Black Rhinoceros was in acute danger in Chete Safari Park as well as in most parts of Africa. I always think that the rhino's motto is 'when in doubt, charge' and it is absolutely tragic to think about this brave animal charging headfirst into deadly automatic rifle fire. Their horns, which are used for defence, are no match for the lethal bullets.

The rhinos were not the only animals in danger, especially on that particular day. Our walking party had unwittingly split up into two groups, ours being the slower of the two. The trek was hot and hard-going. John Kelly, who was suffering from severe cardio-vascular disease and subsequently underwent heart/lung transplant surgery, was making a valiant effort to keep up but, due to his acute health problems, our pace was slower and we fell behind. It was then that I realised the scout with the rifle was leading the other group and our group's leader, ranger Norman English, was armed only with a camera tripod.

The distance between us and the other group widened to the point where we completely lost sight of them, and Norman started to look more than a little concerned. Eventually, he told John and me and the television cameraman to wait exactly where we were and not to move the slightest distance from the tree where he left us while he raced off to find the rest of the group. As we waited for what seemed like an eternity to be 'rescued', I thought about the safety of my Zoo office, my lovely home and garden, my wonderful husband, whether lions and leopards hunted in the daytime and whether they could be frightened off by a small woman wielding a tripod. As we sat in the shade of that security tree I tried not to think about the seriousness of being lost in the African bush. If I did, fleeting feelings of something resembling panic came over me. But any danger we may have been in paled in comparison with what the rhinos faced almost every day.

Finally we heard some faint rustling in the undergrowth and were very relieved to see ranger Norman English, who always went bare-footed, come into the clearing followed by the members of the other group.

I was very glad to see our camp at the end of that long day and John Kelly was, too. I naturally thought it was because he was exhausted from the trek and had been a little nervous about our possible demise. He confessed to me, however, that it was because he was extremely worried about attracting the wrath of one Marguerite Tuit, my personal assistant at the time, who had apparently given John very explicit instructions about looking after me in Africa. Here he was almost losing me in the very first week!

THE TEAM TRACKED and captured ten rhinos while we were in Zimbabwe. The Zimbabwean vets and John Kelly, with his vet's hat on, administered to the local human population as well as the animals in this remote outpost. I saw them treat scouts' cuts and abrasions, listen to their chests and offer advice about clearing up bronchial infections.

A young boy and his grandfather appeared silently out of the scrub one day, seeking help for a wound on the child's foot caused by a bite from a poisonous snake. We sat the child down on an Esky and the vets examined the rather nasty looking wound. It required cleaning and some antibiotic powder, so out came the veterinary kit yet again. Now this kit was equipped to capture the rhinos, so on the top layer of the box were numerous large and particularly serious-looking darts and needles about ten to fifteen centimetres in length, designed to penetrate the thick hide of the animal. The poor little boy took one look at this gruesome sight, leapt up and hobbled off into the bush. While his grandfather endeavoured to encourage him back, we hid the needles and then soothed the nervous little patient with a trusty bottle of Coca-Cola while Mike and John treated his wound.

Our days were long, hot and exhausting but it was an experience I will never forget. Everyone who has travelled to Africa will relate to just how we all fell under the magic spell of that amazing land. John, Mark, Claver and all the Channel 9 gang were fun to be with and I decided I had really won the lottery as far as great travelling companions go. They were always patient, kind and generous, except perhaps when they wouldn't let me buy a two-metre-tall, carved timber giraffe from a roadside market.

We bounced our way in the back of a truck over the roughest roads imaginable following close on the tail of the recovery vehicle. Once the scouts located a rhino they would radio the helicopter, and the pilot would circle the animal. With great skill, vet Mike Kock would dart it from the air and watch where the rhino fell. The scouts and the recovery Uni-Mog truck would then close in and, with a great deal of human effort, move the huge three-tonne rhino onto a stretcher and then onto the tray of the recovery truck. This dangerous work was all done to the sound of a haunting African chant sung by the scouts. I must mention that this truck had three flat tyres on the first day, so you can imagine how rocky and rough the ground was.

Then we all travelled on more rocky tracks and endless

A male lion awaiting instructions from Head Mammals Keeper, Dave Cody, on how best to terrify a young and innocent public relations officer.

In May 1980, Taronga Zoo's Chimpanzee colony took to the trees in their new half-hectare outdoor moated exhibit, which had taken three years to build. It is still one of the finest Chimpanzee displays in the world.

Cotton-top Tamarins, tiny creatures with big hair, share Taronga's Amazonia exhibit with Emperor Tamarins, Squirrel Monkeys and South American Alligators. Tamarins have a 'helper' system when rearing their young, which sees the father and older siblings carrying the babies when they are not being suckled by their mother.

Handsome Archie, a hybrid Orang-utan born at Taronga in May 1975, moved from bars to branches in March 1994 when the Zoo's new multi-million-dollar Orang-utan Rainforest opened.

Crab-eating Macaques meeting — probably to plan future expeditions and outings.

A Taronga-bred rare Sumatran Tiger cub receives his first full veterinary check at six weeks of age.

The undisputed King of the Zoo, Chester the White Tiger, had many fans right across the country. He thought it was all a bit of a yawn.

Minnie the Spider Monkey, everyone's favourite birthday girl, enjoys her sugar-free banana birthday cake complete with carrot candles and walnut decorations.

The elusive and shy Snow Leopard.

One more Black Rhinoceros was added to the world's meagre numbers on 25 May 1996. Kulungwizi, captured in Chete Safari Park, Zimbabwe, in June 1992, gave birth to a male calf at Western Plains Zoo, Dubbo. The calf was named Kusamona (meaning first-born in the Shona language).

Australia's Grass Owl is seldom seen in the wild but Rick's photograph of these chicks, hatched at Taronga Zoo in 1989, was seen around the world.

John Kelly and me smiling for Mark Baker's camera. This photo was taken in the midst of rhino capture action in remote Chete Safari Park on the shores of Lake Kariba, Zimbabwe, in June 1992.

The late Dr John Kelly, Director and Chief Executive Officer of the Zoological Parks Board of New South Wales from 1987 to 1997. His vision and drive were responsible for Taronga and Western Plains Zoos' commitment to rhinoceros conservation.

Buluman, the huge but gentle Western Lowland Gorilla silverback who lived at Taronga Zoo from 1961 to 1980, enjoys an ice cream on a hot summer day.

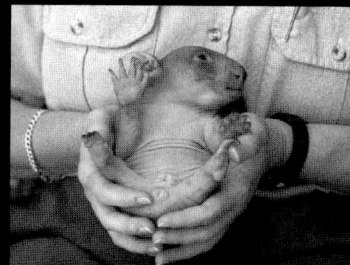

Orphaned baby wombats whose mothers have usually been killed by cars are brought to the Zoo for hand-raising. How could anything this gorgeous end up causing such mayhem?

De Brazza's Monkeys originate in the forests across central Africa. At Taronga Zoo they occupy their own forest habitat alongside the Western Lowland Gorillas. This quaint two some was born in 1992.

unsealed roads to the holding yards or bomas where the vets de-horned the rhino, took measurements, blood and other samples and allocated an identification number to each animal before the tranquillising drug was reversed and the rhinoceros was back up on its feet wondering where on earth it was.

This amazing process was repeated nine times during our stay in Chete Safari Park and the following week the five remaining rhinos were also captured and prepared for their journey to a new life halfway across the world. Everyone involved worked incredibly hard in tough and primitive field conditions and I could only be full of praise for those wildlife conservation 'warriors' of Zimbabwe's National Parks Service.

If we had left these 15 Black Rhinoceros in Chete they would not have survived the year. Over the centuries humans have dealt the rhino a near-fatal blow. It is now time to try and redress the wrongs and bring it back from the edge of extinction.

Animals bring me great joy, but with that comes heartache, too, and none moreso than that associated with the Black Rhinoceros Conservation Project. I try to block out the personal agony many of us felt, especially people like Western Plains Zoo's totally committed veterinarian David Blyde, Director Ian Denney and rhino keeper Phil Whalen when we lost the two male rhinos from Zimbabwe. One adult male succumbed to a liver disease when he was in quarantine on Cocos Island. The other younger male suffered a trauma as he was being unloaded at Western Plains Zoo and died from head injuries that night. Working at a zoo means we are dealing with life, so we deal with death on an almost daily basis as well. This project was always high-risk and John Kelly ensured that everyone involved recognised this and moved on from these great disappointments. We had to learn from them and look to the future with courage and commitment.

DESPITE EARLY SETBACKS, three Black Rhinoceros calves have been born and successfully reared at Western Plains Zoo and as I write this two more rhinos are pregnant.

The first born, subsequently named Kusamona, came into the world on 25 May 1996 and I travelled to Western Plains Zoo that morning to assist with the co-ordination of media coverage of this milestone birth. How excited was I? The sight of the day-old calf brought tears of joy to my eyes. There he was in the Dubbo sunshine — which always reminds me of the clear, bright light in Africa — tottering around on his too-large feet, looking as though he had been born in an old grey overcoat into which he needed to grow. My thoughts immediately went back to Cody, the rhino calf I had met all those years ago on my first day at Taronga Zoo.

Four weeks later, once the new mum and calf were happy and contented with their surroundings and the baby rhino was progressing as it should be, I went back to Western Plains Zoo with a contingent of international, metropolitan and regional media for their first look at the precious new arrival. I think Mark Baker from Reuters was first on the plane to Dubbo that morning. Being present when the mother, Kulungwizi was captured in Chete gave Mark a very personal interest in this particular media call.

Head Keeper Phil Whalen estimated that the calf had put on 30 kilograms since birth and was now a bouncing 60 kilograms. Zoo staff had built a grandstand for the media and when all cameras were focused, Phil gave the signal and out trotted Kulungwizi and her million-dollar baby.

The calf seemed to play directly to the cameras — gambolling, trotting, cantering, falling over, playing with sticks and stones and head-butting his mum. It was a magic moment and the photographs went around the world. How proud were we that we had added one more Black Rhinoceros to the world's meagre numbers.

Kusamona is now over five years old and it is time for him to leave Western Plains Zoo to utilise his valuable genes in the worldwide conservation program for Black Rhinos. He has recently travelled to Fossil Rim Wildlife Centre in Houston, Texas in the United States, with the hope that one day he will sire Black Rhinoceros calves, which may ultimately return to live safely on the shores of Lake Kariba in Zimbabwe. An e-mail received at the Zoo from Bruce Williams, Vice President of Conservation at Fossil Rim, did my old zoo heart a power of good. Bruce wrote that 'Kusamona arrived in wonderful shape and that the entire transfer went like clockwork from transport through to offloading into his new facility.' He described the young male rhino as a wonderful addition to Fossil Rim's group of rhinos and described him as 'a handsome man'.

THE HUMAN DEPRIVATION in Africa was evident everywhere and I thought often while I was there, and many times since, about the slim hope for wildlife survival if the human species faces such overwhelming difficulties. When we were there in 1992 Zimbabwe was probably the most stable and affluent African nation. Sadly, the situation in that country has changed for the worse since and Zimbabwe now must deal with drought, political unrest and economic upheaval. The plight of that country's unique wildlife is now even more perilous.

An oasis for the human animal is provided in that part of Africa by the nuns of The Little Company of Mary in their mission hospital at Murambinda. John Kelly's sister, Helen, was a nursing

Even Greater News for Rhinos

THE ZOOLOGICAL PARKS BOARD of New South Wales has a long-term strategy to be involved in the display, breeding and management of the Black, White and Greater One-horned Rhinoceros and to have one of the most comprehensive rhino breeding complexes in the world.

Taronga and Western Plains Zoos' sister zoo in Japan, Nagoya Higashiyama Zoo, has recently presented a Greater One-horned Rhinoceros to Taronga. This is the first of this species in Australia. After quarantine and acclimatisation at Taronga Zoo, the young male rhino will ultimately live at Western Plains Zoo and plans are underway to bring him into the global breeding and conservation program for this species.

Zoos provide vital gene reservoirs for endangered species while also being able to research aspects of the animal's behaviour and biology, which can help rhinos in the wild.

The Greater One-horned Rhinoceros, which comes from the floodplains and riverine grasslands of Northern India and Southern Nepal, has hide that appears to be rivet-plated and knobbly and, along with the White Rhinoceros of Africa, is the largest land

> mammal, after the elephant. Adults can weigh up to 2,200 kg and grow to 2 m in height.
>
> There are approximately only 2,200 to 2,400 Greater One-horned Rhinoceros in the wild and 134 in the global captive population but there is great hope for this rhino. As a result of huge efforts by the Indian and Nepalese wildlife authorities, this species of rhino's numbers have recovered from under 200 in the early 1900s to around 2,400 today.

nun at the mission and we made a quick visit to her while we were in her part of the world. The mission hospital serves thousands of people from the surrounding rural communities and the nursing nuns, the local staff and volunteers do an amazing job, against incredible odds, to provide much-needed medical and spiritual care.

When we returned briefly to Harare at one time during our visit it was Helen's smiling face that welcomed the weary rhino-catchers at the airport. She came back to our hotel and John invited her to his room for a coffee. Halfway through their chat there was a knock at the door and the hotel manager, a very large black Zimbabwean, strode in announcing that it was hotel policy that no women were to be entertained in male guests' rooms. Poor John got such a fright he declared with complete conviction that 'Helen was not a woman, she was his sister and a nun what's more'! He quickly realised how lame his explanation sounded, apologised to Helen and escorted her down to the hotel cafe to continue their coffee and chat.

Encouraged by my stay at the mission and under the guidance of the nuns, I was able to assist, albeit in a very small way, a local boy, Austin Mutandwa, gain a much-needed secondary education. It was extremely rewarding to see this young boy learn and develop. I am grateful to my friends who also supported me with this project. I also found it especially rewarding working with, St Kevin's Church, Eastwood, in support of the Murambinda Mission Project the parishioners there undertook for some years.

The late Dr John Kelly, who had the determination of a rhino to succeed with this conservation project at Western Plains Zoo — against all odds and with numerous setbacks — still inspires Zoo staff today to continue the work he commenced in the late 1980s. One of the most difficult news releases I had to write during my Zoo days was the announcement of Dr John Kelly's death in November 1997.

John's passing was a tragedy and somehow unexpected by Zoo staff, even though he had been very ill for at least five years. John always bounced back and, to many of us, seemed almost indestructible, a bit like the rhino. I couldn't put any adequate words together for the public announcement of John's death. Mark Williams, my second in charge at the time, wrote the outline of the news release and I sadly added some more personal aspects that, I hope, reflected my own and the rest of the Zoo staff's deep regard for John as a man, as a leader, as a conservationist and as a friend.

Rhino Facts

THE FRONT HORN OF a rhino is the longest and has been recorded measuring up to 1.4 m.

The Black Rhinoceros's skin is actually grey. Its dark appearance is usually as a result of dried mud on the bare skin. It is the most aggressive of the rhino species, although some reports describe them as timid with their charging behaviour designed more to scare off an intruder.

A male rhinoceros's penis faces to the rear so urine sprays out between the back legs.

A charging Black Rhinoceros has been recorded at a speed of 45 kph and even at this high speed it is capable of sudden and fast changes in direction.

The White Rhinoceros is the most social and placid of the rhinos. Mothers and calves can stay together for long periods. Up to seven juveniles form small herds. Mature males tend to be solitary.

All five rhinoceros species are CITES (Convention on International Trade in Endangered Species) listed and five species are protected. Unfortunately, this has proved relatively ineffective and the animals continue to be poached for their horns, which are ground up and mainly used in traditional Eastern medicines as a fever reducer.

In some Middle Eastern countries the horns are carved to make elaborate dagger handles.

CHAPTER 3

Tales of a Monkey or Two

THE MONKEY SPECIES — the primates that mostly have tails — which live at Taronga Zoo change from time to time but have included over the years Mexican Spider Monkeys, Squirrel and Capuchin Monkeys from South America, the Crab-eating Macaques from Southeast Asia, the De Brazza's Guenon from Africa, several species of tiny and exquisite tamarins from South America and the beautiful lemurs from the island of Madagascar.

The Crab-eating Macaques (*Macaca fascicularis*) are small, extremely busy monkeys from the mangrove swamps of Southeast Asia. They used to live in the old monkey pits just up the hill from Taronga's Elephant Temple, and were always up to mischief. I am sorry that the Crab-eating Macaques have now moved on but am happy that their outdated concrete pits have been replaced by the beautiful 'Creatures of the Wollemi' exhibit which opened in December 2000. While this showpiece faithfully replicates the landscape of a Blue Mountains wilderness, I sometimes thought we should have included a cheeky macaque or three in there with the Australian native mammals, birds and reptiles just to liven things up a bit.

It always amazed me how many personal possessions Zoo visitors accidentally dropped into the old monkey pits as they leant over admiring these cheeky, active and inquisitive animals. Lipsticks, hats, mirrors, sunglasses, pens, student work sheets, cardigans and toys of all description found their way into the busy hands of those little monkeys. It was almost a full-time job for the keepers to retrieve the lost items.

My mother, Dora, and sister, Joanne, were in the Zoo one afternoon and came rushing to my office extremely worried that one of the monkeys 'had a machine gun'. I raced up the hill to the monkey pits to find that the macaques had a toy machine gun which one of them had slung over his back by the strap and was chasing the 20-strong troupe of monkeys around, bailing them up at every opportunity.

I don't know which television program this monkey had been watching but he had certainly been to the 'Elliott Ness School of Machine Gun Technique' and he had the other macaques well and truly fooled. Their poor little grey faces looked quite pale with fear. It took a lot of coaxing from their keeper before the gun-toting monkey would hand over his weapon and surrender.

Some of the lost items were never successfully retrieved and I often wondered if the monkeys had some hidden storehouse of goodies they were saving for a giant garage sale or for their glory boxes!

PRIMATE KEEPER JOHN PRINGLE caused me some moments of consternation one day when he came to my office and asked me if I would buy him some lipsticks. Now, I had known John for a

MONKEY FACTS

MONKEYS ARE DIVIDED into prosimians or primitive monkeys — the forerunners of monkeys and apes — and Old World and New World monkeys. The lemur is a prosimian, while tamarins, Squirrel Monkeys and Spider Monkeys are classified as New World monkeys. The Crab-eating Macaque and De Brazza's Guenon are called Old World monkeys.

Monkeys are one of the largest and most diverse groups of mammals. There are over 240 species spread across Asia, Africa and Central and South America. Many species are endangered by loss of habitat.

Monkeys are intelligent animals, quick to learn, inquisitive and have an excellent memory.

They range in size from the large and boldly colourful Mandrill (*Mandrillus sphinx*), which comes from the western central African coast and weighs in at around 37 kg to the tiny Pygmy Marmoset (*Callithrix pygmaea*) found in the forests of western South America. The Pygmy Marmoset is the world's smallest monkey and at a diminutive 125 g it could fit into the palm of a human hand.

The strangely beautiful Snub-nosed Monkey (*Rhinopithecus roxellana*) survives in the cold mountain forests of eastern Asia where it endures average winter temperatures of −5°C. At the other end of the temperature scale the strange-looking Proboscis

> Monkey (*Nasalis larvatus*) lives in habitats near water in the hot and steamy lowland rainforests of coastal Borneo.
>
> With all monkeys, relationships are generally very close and grooming is a mutually beneficial and significant social activity.
>
> They nearly all have long tails. However, in some monkey species the tails are underdeveloped but their tales never are!

few years, not overly well, but I didn't think he had a penchant for women's make-up.

I asked him what colour he wanted and he responded, 'Just get a colour the macaques would like'! The Mosman chemist shop assistant looked at me very strangely while I tried numerous colours in her range of lipsticks before I selected several that would be attractive to the monkeys.

I delivered the lipsticks to the monkey keeper who then, with some relief on my part, explained his need for the make-up in greater detail. Certain Crab-eating Macaques were proficient escape artists who regularly scaled the walls of their exhibit to explore the surrounding Zoo gardens. They had become somewhat dangerous as they were inclined to run up to visitors and grab their ice creams or whatever they happened to be eating. While they look innocent enough, monkeys have sharp nails and large, dirty teeth and a bite or scratch from one would be serious.

When the keepers approached the escapee macaques in an effort to catch them, they quickly ran back into their pit and blended in

The Customer is Always Right

I SHOULD ADD HERE that the same chemist shop assistant looked equally aghast at my next request a few months later when we had to buy cough mixture for the fluey chimpanzees. I needed 'a litre of sweet-tasting cough syrup suitable for ages from four months to 40 years'!

with all the stay-at-home monkeys, making it difficult for the keepers to identify them and do anything about the problem.

The macaques loved playing with lipsticks that were dropped into the exhibit by visitors. They managed to get lipstick not only on their lips, but all over their little faces and hands as well. Keeper John Pringle's theory was that if I handed one of the escapee monkeys a lipstick while it was out of the exhibit, it and the other truants would almost certainly get covered in it. When they eventually ran back into the group the keepers could identify the culprits because of their bright lipstick stains.

So there I was armed with my lipstick lures, which the macaque escapees duly grabbed from me while John and the other monkey keepers kept out of sight. The monkeys snatched them and ran off, but instead of plastering themselves with colour they removed the cases and ate the contents whole! Not a lipstick smudge to be seen.

The next day we tried again but this time John armed me with syringes of red cochineal, which I squirted at the truants.

I felt like a traitor as the red-stained culprits were caught up later and shipped off to a new, more secure home in a north Queensland wildlife park.

SOMEONE ELSE IN THE Zoo must have had the same thoughts as me about macaques livening up other exhibits because when the new Orang-utan Rainforest was designed it included a place for macaques to live as well. The Crab-eating Macaque is Southeast Asia's most common monkey and in the wilds of Borneo and Sumatra these two species would co-habit, so displaying them together at the Zoo would provide an even more realistic picture of rainforest life. Unless, of course, the monkeys decided they would like to take a ride on the Sky Safari cable car, not many of which are to be found in Southeast Asian jungles.

The Sky Safari gondolas travel from the ferry wharf, up the steep slope of the Taronga site to the top of the Zoo and back again. This exciting new transport service was first introduced to Taronga in 1987 and then substantially upgraded and re-opened in May 2000. The six-person high-speed cabins glide just above treetop level on a cable strung from tower to tower and Zoo visitors can enjoy fabulous views of the city, the harbour, a bird's-eye view of Taronga and sometimes come face to face with a macaque or two.

As I have mentioned before, Crab-eating Macaques are expert climbers and seemed to spend a good deal of their time planning outings and excursions into the Zoo grounds. They generally never go too far from their exhibit but usually manage to cause chaos nevertheless.

Under the weight of heavy rain, overhanging branches were now too close to their exhibit walls, giving the ever-curious and constantly plotting macaques a rapid and undetected escape route. From these branches the macaques swung onto one of the Sky Safari towers and then hoisted themselves up onto the cable, swinging like skinny, furry Tarzans, hand over hand towards an oncoming gondola.

This was not looking good. The looming gondola seemed huge and threatening and the macaques looked very small and helpless. I found myself starting to feel sorry for the monkeys but that didn't last long — I should have known better than to waste my pity on these wily little creatures.

The code for an animal escape went out over the two-way radio and keepers came from all points of the Zoo to help. When the problem is such a long way off the ground it is very difficult to provide instant assistance. There wasn't even time to stop the Sky Safari.

There was much wringing of hands and several tearful keepers crying: 'What will we do? What will become of the macaques?' John West, Taronga Zoo's senior Operations Manager and a veteran observer of countless monkey escapades, was heard to mutter: 'Don't panic, they are bloody monkeys, they'll work it out.'

And, of course, John was right. Just as the gondola was almost on top of them and I almost didn't dare look for fear of seeing Tarzan and friends squashed flat, the Crab-eating Macaques did lightning fast sideways leaps, landing on the tower to wait until the cabin had passed by. They then jumped straight back on to the cable for some more fun and games.

Ultimately, the monkeys tired of this escapade, returned to their exhibit and the offending branches were trimmed. I am sure there was a lot of chatter in the dens that night as the macaques tried to interest the slow-moving orang-utans in possibly joining them on their next Sky Safari excursion — maybe on a one-way ride down to the ferry. A fleeting image of these escapees, not only expert trapeze artists but also extremely adept swimmers, foregoing the ferry and diving into the harbour on their next outing didn't bear thinking about even for an instant.

OVER THE YEARS the macaques' Houdini-like escapades have tested keepers' patience and caused keepers' blood pressure to rise when their antics have been substantially misreported.

In the early days before two-way radios for staff were introduced in the late 1980s, the only means of urgent communication was to the central Zoo switchboard via the two or three internal telephones, situated at specific behind-the-scenes locations around the grounds. If you had a problem, invariably the nearest telephone was some distance away, which meant a mad dash through the grounds often resulting in a somewhat breathless message being relayed at the end of it.

One morning a junior monkey keeper — who shall remain nameless — arrived for work and noticed that the macaques had, yet again, gone on one of their nocturnal excursions. Several of them were in nearby trees peering down at him with sheepish grins and a good deal of blinking which in 'macaque-speak' says something like: 'I'm really big and tough, so keep away.' Desperate for some assistance to catch them up before they got into too much mischief, the keeper ran to a telephone at top speed and reported, somewhat breathlessly to the switchboard operator: 'The macaques are out! The macaques are out!'

Unfortunately, the switchboard operator thought this gasped and somewhat garbled message was a zookeeper's worst nightmare and urgently passed it on to the Head Keeper as: 'The big cats are out, the big cats are out.' You can imagine the activity that that little gem caused.

Once the tumult and shouting had died down and it was discovered that much had been lost in the transfer of the message, the macaques were duly caught up yet again and marched back to their exhibit.

ONE OF MY OTHER favourite monkey tales involves the small, brown and extremely adept Capuchin Monkey (*Cebus apella*). This monkey comes from South America, where they live in the tops of the large rainforest trees from Honduras in the north to Argentina, and in the south, just east of the Andes Mountains. The Capuchins are often regarded as the most dextrous of New World monkeys. They eat fruit, leaves, shoots, insects, eggs and young birds and have been observed rhythmically hammering open hard objects in their search for food.

Taronga displayed the Brown Capuchin, which is also known as the Tufted or Black-capped Capuchin, since 1936. The last ones left the Zoo in 1988.

Monkeys are such clever creatures — their keepers continually need to devise activities that enrich the animals' behaviour and keep them stimulated. Primate Division Manager Paul Davies, who had one of the most challenging jobs in the Zoo, came up with the bright idea of providing his Capuchins with their own food dispenser so they could learn to twist a knob and access dried fruits and nuts for themselves. He bought a large and sturdy clear Perspex lolly dispenser, filled it with the monkeys' favourite treats and put it in the exhibit. I invited the media in for a photo opportunity and we all settled back to watch the monkeys have some fun.

They were intrigued by this new 'toy' and played with it enthusiastically, quickly working out that by twisting the knob they were rewarded with a delicious peanut or raisin. It was wonderful to watch the Capuchins thoroughly enjoying this new activity and as usual the media left the Zoo with some great photographs. I didn't dare tell them that by the very next day the clever Capuchins had dispensed with twisting the knob to retrieve a rather boring single nut. They knew exactly how to unscrew the wired-on lid and were helping themselves to fruit and nuts by the handfuls! Needless to say, the dispenser was soon removed.

MINNIE WAS A MEXICAN or Central American Spider Monkey (*Ateles geoffroyi*) who broke all longevity records Minnie, who had come to Taronga as a youngster in May 1949, was the one animal in the Zoo who we indulged, in her later years, with an annual birthday party. Her keepers, firstly Suzie Haddock, who handed her special love of Minnie (and her recipe) on to keepers like Glen Sullivan, Debby Cox and Lisa Abra, made a sugar-free carrot cake for her, iced it with mashed bananas and used carrots for the many birthday candles.

The party guests were usually local Mosman school children or residents of the nearby retirement village. We all sang 'Happy birthday, dear Minnie' at the top of our voices as she and the other Spider Monkeys tucked into the cake. The media always turned out for Minnie's parties and I think she really enjoyed being the centre of attention as her rheumy old eyes seemed to

> ## SPIDER MONKEY FACTS
>
> SPIDER MONKEYS COME FROM the forests of Central America and Mexico where they live in social groups. They have a thin body, a prehensile tail, long limbs and thumbless hands which act as simple hooks enabling the monkeys to swing agilely through the trees or to pull fruit-laden branches to their mouths.

light up for the cameras. Minnie's birthday snaps are among my favourite press photographs.

Although Minnie loved the cake, this old monkey very obviously didn't like sticky hands and as soon as the birthday cake was finished, Minnie would be seen down at the water trough or moat washing her spindly fingers and drying them carefully on her furry backside.

Minnie lived her long life in the same enclosure at Taronga, just down the hill from the main Zoo cafe and adjacent to the famous Floral Clock. Zoo archives show that Albert Le Souef, the Director of the old Zoo at Billy Goat Swamp or Moore Park, which is now the site of the Sydney Girls High School, was very much involved in the drawing up of the master plan for the new Taronga Zoo. The earliest drawings and specifications for the Zoo at Mosman show 'provision for ten monkey enclosures' and one of these would seem to have been what is now the Spider Monkey enclosure, as archives indicated that Spider Monkeys have been displayed at Taronga since around 1918 and that there was at one time a Spider Monkey called Mary who 'was then the oldest mammal in the Zoo'. The original enclosure was a completely circular structure with viewing from all sides. Later, probably about the time Minnie arrived at Taronga, the southern side of the exhibit was walled and a half-canopy was added to give the monkeys protection from the cold winds and rain. It was always a high profile location and a popular stopoff point in a Zoo visit. Many fans regularly called by to greet the popular Minnie and her predecessors. One day, however, Minnie had the opportunity to step outside her domain and greet them.

A distracted keeper had left a bolt on a door undone long enough for a wily old monkey to creep out and head for the cafeteria, perhaps to see if there was any birthday cake on the menu that day. By the time her escape had been discovered, Minnie had checked out the cake-less cafeteria and had decided it was not worth leaving home for. She did, however, have her eye on a schoolboy with an interesting looking ice cream in his hand. Rather than have to net or dart the old monkey, Dave Cody and Graham Button asked the rather amazed child to slowly make his way down to the Spider Monkey exhibit in the hope that Minnie would follow the little boy and the coveted ice cream.

When the other Spider Monkeys saw Minnie strolling down the path towards them they started chattering at the top of their little voices. Minnie looked quite relieved to see their familiar faces, forgot the ice cream and quickly went to the door to be let in. It was the first and last tour abroad for this long-time Taronga resident but I am sure she thought and talked about it until the day she died.

The Spider Monkeys often made me giggle and cheered up my early morning activities in the Zoo grounds especially on the dark, cold days that sometimes started for me before the sun had come up. I regularly needed to meet radio or television

Please do not Feed the Animals

LONG GONE ARE the days when visitors to Taronga could buy bags of broken biscuits to feed to the Zoo animals. This was a tradition at Taronga in the 1930s and 1940s. Now, there is a strict 'Please do not feed the Animals' policy and for very good reason. All the animals are on specially prepared and scientifically devised diets to maintain good health and avoid obesity.

Primates in particular are susceptible to many human diseases such as colds, influenza, herpes and tuberculosis. Even a half-eaten apple innocently thrown to the monkeys by a visitor with a cold can cause illness for the animals. It can also cause fighting and injury.

The feed bill for Taronga and Western Plains Zoos comes to around $800,000 annually and requires the food preparation officers to source and purchase everything from mealworms to mung beans, from maize to mullet and everything in between.

technicians, sometimes as early as 4.00 a.m. and give them access to the Zoo to set up for live broadcasts. Invariably, as we walked quietly past the still sleeping Spider Monkeys on our way to the broadcast location, Minnie and her ever-alert troupe would wake up and start abusing us loudly. Their chatter, which sounded like a string of what would be the equivalent of monkey swear words, would follow us until we were out of earshot, and always made the early start worthwhile.

OVER MY ZOO YEARS, I received very few scratches or bites from the animals. In fact, it was probably a few of the stranger human animals I encountered that had the real teeth and claws! Old Minnie, however, landed a real left hook, which would have done a world boxing champion proud.

Towards the end of her life she had a bad dose of influenza one winter and was recuperating in the holding yards at the rear of the veterinary hospital. I was taking a shortcut to the Bird House one morning and noticed old Minnie on her own, looking quite forlorn. As she was basking in the winter sun I went up to her cage and encouraged her to come over. Hoping no one was watching, I got close and quietly asked her how she was feeling.

As quick as lightning her black, scrawny hand shot out through the wire and slapped me with the most amazing strength, right on the bridge of my nose. Ouch! I saw green stars for quite some minutes but kept the mishap to myself, as I really should have known better than to get that close to a clever and cunning little monkey. Minnie may have looked old and frail but she could still pack a punch.

It was like losing a treasured member of the family when old Minnie passed away in 1993 aged approximately 45 years. Even though she had broken all longevity records for Mexican Spider Monkeys, it was a bleak day indeed.

CHAPTER 4

Big Cats and Small

TARONGA'S HEAD MAMMALS KEEPER Dave Cody was always a very mischievous man. I think he caught it from the monkeys at the Zoo which he had been hanging about with for over 30 years. He loved playing tricks on people, especially young, inexperienced public relations officers. I was to discover this very early in my Zoo life when he invited me to go with him on his 'unofficial', behind-the-scenes tour of the lion exhibit.

African Lions (*Panthera leo*) have been displayed at Taronga since the Zoo opened in 1916. The early lion and tiger exhibits — which had moats rather than bars separating the animals from the visitors — had been very progressive and innovative for their time. Those big cat exhibits, following many and varied makeovers in the past 85 years, are now the location of Kodiak Bear Canyon and home for bears Barney, Bethel and Cynthia.

The lions — big cats synonymous with the African savannah — were later displayed in deep concrete pits and spent their days lazing in the sun on mock rock perched on top of a concrete base in the shape of a map of Australia. Even as a child visiting Taronga, I could never work out why this magnificent King of the Beasts should be displayed this way, or what on earth lions had to do with our fair land. Fortunately, this exhibit was changed substantially in the late 1970s and again in the 1990s to something more in keeping with the lion's natural habitat.

As I innocently followed Dave into the back of the old lions' den I obeyed his every instruction. He told me that whatever I did I must stay away from the den doors on the left and keep very close to the wall and dens on the right. The den on the left contained, according to Dave, a very cranky hyena, which apparently had a nasty habit of snapping at anyone who came too close.

It was quite creepy in this cavernous dungeon, which had lots of spider webs, dark corners and an eye-watering smell of very strong cat urine. I was careful to do exactly what Dave Cody told me as I followed him in. In fact, the right-hand wall and I were as one as I pressed myself up against it to avoid the reputedly dangerous creature lurking across the alleyway. The next thing I knew, a huge male lion weighing about 250 kilograms — the rather angry resident of the den on the right — had lunged at the bars roaring at the top of his lungs as only an adult lion can. He practically seared me with his hot, smelly breath, and nearly burst my eardrum — not to mention almost changing the colour of my underpants!

I ran shrieking from the scene as Dave convulsed with laughter at the success of his trick. I was to discover that this was one of Dave's favourite 'induction' tricks for all newcomers to the Zoo and that he even tried it on some of the more elderly Zoo Volunteers until he was 'advised' to cease and desist: I also learned that the much-feared hyena opposite was, in fact, a geriatric creature incapable of snapping at anything more than a fly.

These Cody tricks were all prior to the advent of occupational health and safety regulations, which with good reason have been embraced with appropriate commitment by the Zoos. These changes have probably meant, however, that zookeepers today do not have the same fun that Dave Cody enjoyed.

From this experience, it is difficult to understand how I could possibly have come to enjoy such a strong friendship with Dave over the years or how I developed such a passion for all things feline. Both certainly got off to a shaky start. I greatly missed the mischievous Dave once he retired in 1988. The Zoo seemed a very 'tame' place after that.

BIG CATS AND SMALL give me great joy. Everything from moggies to magnificent Sumatran Tigers make me purr with delight. Rob and I have shared our lives, lounges and our garden with four strays found in the Taronga car park over the years. It is not so common now, thankfully, but when I first started at the Zoo people were dumping cats in the car park or over the Zoo wall, apparently thinking that we would like to have them. The last thing the Zoo wants is stray cats prowling around carrying disease and attacking and decimating the large population of free-ranging bird life. These cats were caught up and, if homes were not found, would be humanely put down.

The keeper in charge of the Veterinary Quarantine Centre Tony Carrick was a soft touch and went to great lengths always to find suitable homes for the car park strays. I ended up taking home two ginger toms whom we christened Rudolf Nureyev and Mitka, a white female subsequently named Anna Pavlova and her grey

LION FACTS

THE AFRICAN LION (*Panthera leo*) is distributed over central and southern Africa where it lives in open savannah and plains country. It is reduced in range now due to alteration of its environment and habitat.

This carnivore feeds on a diet of grazing animals such as zebra and wildebeest.

The social lions live in family groups, called prides, which usually consist of a dominant male, several females and their cubs. The permanent core of the pride is a group of lionesses and the only really close social bond in the pride is that which exists between a lioness and her cubs. Males are relatively peripheral and short-lived members of the pride and pair-bonding is non-existent.

Lions generally stalk their prey at night, sometimes working in groups. By day, they are generally inactive and like most cats (including the household moggie) spend up to 20 hours a day sleeping.

'beach ball' shaped sister who we called Giselle (certainly no Queen of the Willis).

Mitka Clements first made his presence known at a Zoo Volunteers' Thank You Party at Taronga's Education Centre Christmas 1979. We had set out the food on long trestle tables covered with cloths and staff were mingling with the Zoo

Volunteers when suddenly I noticed a small ginger paw striking out from under the cloth, hooking a chicken wing off a plate. The mother cat and her mischievous ginger kitten made their escape that night but both cats were caught the next day in the car park by Elvine Thomas, fleet-footed wife of Australian mammals keeper David Thomas. Tony used his winning ways to convince me I needed another ginger cat in my life to join the already ensconced Rudolf Nureyev and some other Zoo softie took the mother cat home.

Mitka was de-sexed a couple of months later by the Zoo vet in his morning tea break. Much to my horror, when I came back to my office after a meeting, a little jar was sitting on my desk containing two former bits of Mitka's anatomy. Zookeeper weird sense of humour at its worst! I quickly phoned the veterinary hospital to see how my ginger chap was getting along and was told by Tony that the kitten had 'hired a hospital TV and was sitting up in bed watching the "Midday Show"'!

Mitka reached the ripe old age of 21 and now, along with the other Clemcats, is buried in a sunny corner of our garden. Not a bad innings for a car park stray. I must say, however, that despite my love of cats, I doubt if I would ever have one as a pet again. The native birds, lazy Blue Tongue Lizards, possums and now even a bandicoot, which have come to our garden since the demise of our moggies, are very welcome replacements. It is also nice to be able to provide an occasional holiday camp for a variety of cats while their owners are ill or on holidays.

Not long after I began working at Taronga I read a book by ex-pat Australian John Rendall called *A Lion Called Christian*. John, now the social editor for *Hello!* magazine and a hardworking member of the George Adamson Wildlife Trust, wrote this delightful story of a lion cub he and a friend bought in Harrods department store in London and subsequently reared in their antique shop in Chelsea until it began frightening off the customers. Christian, when almost fully grown, was rehabilitated by the incredible George Adamson and released into the wild in Kenya. It was this magic story which encouraged my early fascination with big cats. This fascination was fed over and over again at Taronga and Western Plains Zoos.

It is not easy to display big cats well in a suburban zoo situation. These animals, fascinating and majestic as they are, spend around 20 hours of their day sleeping, which is usually very frustrating for the majority of Zoo visitors. The people who own a pet cat can sort of understand it. The cat exhibits at Taronga have undergone great changes over the years in efforts to provide improved conditions for the animals and an opportunity for Zoo visitors to view them in the best possible way. They have gone from concrete pits and long, narrow concrete corridors to lush jungle, savannah or mountainous habitats where the animals are viewed through glass, enabling visitors to realise the majesty of these amazing animals. I still feel more can be done to ensure the powerful presence of the big cat is fully appreciated.

The carnivore keepers, the Zoo's behavioural biologist Margaret Hawkins, and her dedicated Zoo Friends Volunteers spend hours devising environmental enrichment programs for the cats. These involve hiding food so the carnivores have to 'hunt' for it, putting down obstacles like logs and branches across pacing paths, placing strong smelling herbs in the exhibits and providing a diverse diet ranging from fully furred rabbits to skulls and bones or tying meat to poles so the cats use their amazing

The Lion's Share

TARONGA'S THREE AFRICAN LIONS eat 3 kg each per day. Monday it's chicken (feathers and all), Tuesday it is horse meat and liver, Wednesday it's half rations (1.5 kg), Thursday it's beef day, Friday it's horse meat again, Saturday it's beef again and Sunday it's kangaroo. All the meat is sprinkled with a multi-vitamin powder.

climbing abilities to retrieve their prize. It's all about keeping the animals busy and interested, making them use their well-developed senses and dexterity to locate and outsmart their 'prey'. After all, in the Zoo, these hunters, who in the wild would often need to cover kilometres looking for food or a mate, have 'meals on wheels' and 'computer dating'.

THERE IS SOMETHING special about cheetahs that really intrigues me. Their amazing acceleration and speed, beautiful eyes, black 'patent leather' lips, long pencil legs and sleek spotted coats epitomise for me feline elegance. Two young cheetahs, on their way to Western Plains Zoo from a wildlife park in England in 1977, stopped off at Taronga to fulfil their quarantine requirements. These two were the spark that ignited my interest in these glorious spotted cats.

The two new cheetah were displayed for about three months at

CHEETAH FACTS

THE NAME CHEETAH (*Acinonyx jubatus*) derives from a Hindi word 'chita' which means 'spotted one' and this also suggests that the species once ranged far further than Africa and were found in northwestern India, Pakistan, Afghanistan, the Middle East and Iran.

The cheetah is a branch of the cat family that specialises in running to secure its prey. This specialisation as runners has called for some modifications in nature's usual cat design. They have longer legs and long, supple backs to allow the flexing and large strides which enable the cheetah to cover ground at up to 112 km an hour over short distances. The speed has apparently been measured accurately on a greyhound racing track with a Cheetah chasing the artificial hare. Their skulls are lighter than most cat species in the interests of balance and speed, and hold smaller, lighter teeth. Their long tails also assist their balance. Their claws are not fully retractable and appear more dog-like than cat-like, acting like the sprigs on running shoes.

Sydney's zoo and then travelled by road to Western Plains Zoo. I accompanied them on this journey and organised media interviews for the vet and their keeper along the way over the Blue Mountains, and in the towns of Lithgow and Bathurst.

Spot the Cheetah

THE CHEETAH IS CLASSIFIED as Vulnerable. There are thought to be between 5,000 and 9,000 left in the southern African countries. Sadly, in Iran where only 40 to 50 animals remain in the wild, the cheetah is classified as Critically Endangered.

There are 34 Cheetah in Australasian zoos. John Lemon the Species Co-ordinator at Western Plains Zoo and Studbook Keeper for Cheetah, looks into their genetic backgrounds, records their histories and spreads their valuable genes in the most effective way to ensure successful breeding.

Western Plains Zoo and other Australasian zoos are involved in the international cheetah breeding program in conjunction with wildlife organisations in Europe and the United States.

Excellent in-situ work is being done by the Cheetah Conservation Fund in Namibia, the DeWildt Cheetah Breeding Centre and by Africat. These organisations carry out rehabilitation and release programs for captive-bred cheetah, remove and translocate problem cheetah and educate local communities on cheetah conservation.

It was obvious from the considerable media and community interest that I was not the only person fascinated by these fastest of all terrestrial animals.

Another cheetah I was honoured to know some time later was born at the DeWildt Cheetah Breeding Centre in South Africa in 1983 and was named Augustus. He arrived in Dubbo in 1985 via Melbourne Zoo and became the sire of four successful litters of cheetah at Western Plains Zoo. Gus, as he was affectionately known, was a particularly quiet and gentle animal who won many a heart. He lived to the grand age of 14 and over those years zookeeper John Lemon often delighted Zoo VIPs by allowing them to go into Gus's enclosure for an up close and personal encounter with this friendly old cat.

I was at Western Plains with photographer Joe Larrarte doing staff shots for the Zoo annual report in 1997 when John Lemon was amazed to discover that I had never patted old Gus. Imagine my delight when he invited Joe and myself to accompany him on his afternoon lock-up round to meet Gus up close. We walked into Gus's domain and waited while John softly called his name. Gus silently crept out of the undergrowth and immediately began circling me, seeming extremely interested in my long woollen overcoat. He repeatedly sniffed and head-rubbed me, all the while giving muffled growl-like purrs, chirps and even a spit and a hiss or two. I saw Joe looking somewhat nervous and I, too, started to feel a little uncomfortable at all the attention. I was just about to suggest we say farewell when Gus came up quickly from behind and bit me on my bottom! Fortunately I was wearing thick winter slacks and my overcoat to protect me, but there were definite teeth marks and substantial bruising there for some days after.

I was mortified that gentle Gus who had never harmed anyone before had chosen to bite me, the woman who was so passionate about cats. John Lemon insisted it was a 'love bite' and thought it was a huge joke. He immediately told the whole Zoo staff what had happened via the two-way radio. By the time I got back to the office everyone wanted to see the damage.

A short while later an explanation for Gus's out of character behaviour became a little clearer to me when I donated a ground-cover conifer from my own garden to Taronga's Snow Leopards. We had planted the conifer too close to a pathway at home and over the years it had crowded out everything surrounding it. Asian Division Manager David Pepper-Edwards gladly accepted my offer of this shrub for Snow Leopard Mountain and it was transplanted by the Zoo horticulturists and placed in the exhibit beside the Snow Leopards' pond. The strong, virile plant lasted less than a week as the two Snow Leopards took an instant dislike to it and stalked it, pounced on it, rolled in it, chewed it and dug it up by the paw-full.

It suddenly dawned on me that for 15 years my cats had peed in this plant, rolled on it and dug under it and the Snow Leopards were taking great umbrage at having their domain invaded by the smell of these suburban moggies. It also dawned on me that old Gus the cheetah had probably picked up the smell of Rudolf, Mitka, Anna Pavlova and Giselle on my woollen overcoat — although I hope they had never peed on it — and his attack on my rear end was somehow explained.

Cheetah breeding has been a success story at Western Plains Zoo and just recently two more male and two additional female cheetah cubs, also from the DeWildt Centre in South Africa, spent a few months at Taronga before travelling to Western Plains Zoo. Hopefully they will continue the breeding successes of the past and follow in the silent footsteps of gentle old Gus.

TARONGA ACQUIRED SOME other equally beautiful spotted cats in 1990 and the new Snow Leopard Mountain was opened with great fanfare at the Zoo in May that year. It was the first new exhibit at Taronga since the Giant Pandas in 1988. It is a very authentic re-creation of a Himalayan mountain habitat with rocks, alpine plants, a stream and pool. Visitors view the beautiful Snow Leopards, a species new to Taronga, through tightly strung piano wire and are led on a journey of discovery through alpine trees and shrubs to the exhibit past tactile sandstone sculptures and dramatic graphics depicting the special adaptations this elusive and shy cat has to ensure it survives in its harsh and wild terrain. These animals, their exhibit and the surrounding alpine meadow are among my Zoo favourites.

For the new exhibit opening ceremony our neighbours, the 10th Terminal Regiment, provided a guard of honour and became the protectors of the Snow Leopards. I also secured the services of a Tibetan lama to perform a long-life and fertility blessing for our rare Snow Leopards. We even managed to find a traditional Tibetan flautist to fill the Zoo with haunting folk music. Even though the day was cold and drizzly, I felt that the weather was in step with the occasion and that of the Snow Leopards' home range.

The lama or monk arrived wearing his splendid saffron robes and as I escorted him through the Zoo past the central seal pools, the seals jumped up at the wire and starting barking what I thought was an excited welcome to this holy man. I was amazed he had such an attraction for the seals and when the opening ceremony was over I told the seal keepers about the powers of the lama and how fascinated the seals were by him. The keepers looked at me very sceptically and asked if he was, by any chance,

SNOW LEOPARD FACTS

THE SNOW LEOPARD (*Panthera uncia*) originates in the high and remote mountains that make up the Himalayas, the Karakorums and the Hindu Kush where they survive fierce winds, thin air and extremes of heat and cold. Sadly this legendary species, which is now classified as Endangered, is having difficulty surviving as it is hunted for its soft, thick fur. The Snow Leopard is also competing with the ever-burgeoning human population, which is destroying its habitat and killing the wild goats and sheep, which are its main food source. The ravages of war would be taking their toll, too.

The Snow Leopard, of all the cats, has the thickest fur and the longest, thickest tail. The tail length is almost one metre and it is used as a blanket, especially to wrap around newborn cubs. It also acts as a rudder as the agile cat leaps about the mountainous ridges above the tree line in its icy mountain terrain. The summers in the rocky valleys can also be very hot so the Snow Leopard moults to keep itself cool. They are generally solitary animals but during the mating season from January to March males and females will bond briefly, hunt together and then separate. One to three cubs are born after a three-month pregnancy and the cubs stay with the mother for the first winter after which they leave to establish their own territory.

wearing yellow? Of course he was. Paul Hare and the other marine mammals keepers shot me down yet again by saying the seals probably thought the lama was a keeper in a yellow raincoat delivering their bucket of fish on a wet day.

The Taronga Snow Leopards, firstly Mangar and Shimbu on loan from Melbourne Zoo and now Omaha, named after the city zoo in which he was born and Profula, Nepalese for 'happy, smiling face', have enjoyed a long and healthy life. Sadly, though, Profula and Omaha have not produced any cubs to add to the world numbers of this shy and highly endangered cat. I am convinced, however, that one day the lama's blessing will be fulfilled and some tiny spotted 'Snowies' will be born in Mosman.

SOME OF THE UNUSUAL SMALLER cats that were displayed at Taronga during the 1970s and 1980s were fascinating but elusive and their shy behaviour often meant they were not the most successful zoo display animals. Two of them, the Jaguarundi (*Felis yagouaroundi*), a small, stoutish felid from southern United States and South America and the Caracal (*Felis caracal*), a very beautiful desert-dwelling small cat with tufted ears which comes from Africa and southern Asia, acquired rather unique aliases. Choro and Conchita Jaguarundi arrived from Rotterdam Zoo in 1979 and Claudius and Ophelia Caracal came the same year from Melbourne Zoo and we were able to generate substantial media interest in these newcomers.

Ken James, a photographer from the *Manly Daily*, arrived at the Public Relations office one afternoon very eager to photograph the new 'Jacarandas' and the 'Caraculars'. I had to think hard about that for a while but soon worked out it was the

Jagurundis and Caracals he was interested in. I am afraid their aliases stuck with them for a long time and I still think of them as that with great affection.

WORLD BOXING CHAMPION Mohammed Ali visited Taronga with a huge fanfare and a monstrous media following in 1979. He does not know it but he was the inspiration for a new Zoo animal's name. Earlier in the week a young black jaguar (*Panthera onca*) had arrived from Rotterdam Zoo and was subsequently named Ali after the Zoo's recent famous visitor. The jaguar, a big cat that originates in Central and South America, is a powerful, heavily muscled animal. The black jaguar is a highly pigmented spotted jaguar. Ali was a sleek, healthy black cub who grew up to be as handsome and athletic as his namesake. He died from old age in 1996 but I should note that Ali sired two offspring, Quito in 1984 and Maya in 1987.

THE 5TH BATTALION Royal Australian Regiment, based first at Randwick and then Holsworthy in Western Sydney, was known as the 'Tiger' Battalion. Taronga was approached about the possibility of them having a Zoo tiger cub as a mascot. This Battalion, subsequently known as the 5th/7th following amalgamation with the 7th Battalion, had a tour of duty in Vietnam and a letter to the Zoo from the commanding officer during that time states 'there were times when we would have appreciated his ferocious presence'. On 12 May 1967, Private Quintus TF005 became the official mascot of this 'Tiger'

TIGER FACTS

THE MAGNIFICENT and charismatic tiger (*Panthera tigris*) has been exhibited at Taronga since the Zoo opened its gates in October 1916. Even though these were probably hybrid animals (a cross between Bengal and Indo-Chinese sub-species most likely) these magnificent orange-and-black cats were superb ambassadors for all tigers, which are tragically disappearing rapidly in the wild.

Of the eight tiger sub-species, the Bengal is the most common but still only 3,000 to 4,000 of these animals exist in India and surrounding areas. The Indo-Chinese sub-species has about 1,500 animals remaining in Burma and Malaysia, while only approximately 500 Sumatran Tigers still exist in the wild. The Balinese, Javan and Caspian Tigers are most certainly extinct, the South-Chinese Tiger numbers are extremely low and it won't survive, and only 300 Amur Tigers remain.

It is the same old story which is causing their demise — destruction of their forest habitat, poaching for their skins and body parts and intense competition from the human animal. National parks are being established as designated tiger reserves, hopefully free from poachers, but more urgent work needs to be done for tiger conservation. Zoos and Zoo visitors can assist this work by supporting conservation initiatives and Taronga Zoo now regularly donates funds for Sumatran Tiger conservation projects in the wild.

Battalion. The Battalion's commanding officer hoped the cub could make parade ground appearances on ceremonial occasions, all the while living at Taronga, of course.

This tiger was named Quintus and he became as famous as his battalion as he attended a variety of regimental ceremonial occasions. He didn't always behave with impeccable parade ground manners. One newspaper cutting reports 'Quintus lay down during the entire parade, even when the Prime Minister inspected the troops'!

As most tiger cubs do, Quintus soon grew into a healthy, strong adult tiger unable to fit into the pet pack which took him to the ceremonial parades. So as each little Quintus grew into a large Quintus he was replaced, where possible, by a cub from a subsequent litter, which were all sired by the original Quintus. When there were no cubs available, the Army even provided a custom-built transport crate in which the original and by-then very large Quintus could do 'tours of duty' at very important parades and ensure that all attending were safe from attack. His parade ground manners must have improved because his records note 'He is an impressive and loyal soldier' and in 1980 he received his Warrant.

Original Quintus was a prolific breeder and became the highest-ranking mascot in the Australian Army when he attained the rank of Warrant Officer Class II, purely, it seemed, on his ability to sire cubs. Devoted Quintus fan and former *Herald* journalist Elizabeth Brown, who closely followed and reported on Quintus's career, once enquired if Quintus was ever likely to be commissioned. The commanding officer was very definite in his response: 'Definitely not, ma'am! We could not have the men saluting a tiger.'

The association with the 5th/7th Battalion RAR was a long and very happy one for Taronga. The 'Tiger' Battalion was always the first to answer the Zoo's calls for assistance with a guard of honour for special exhibit openings or royal or vice-regal visits. It also supported Quintus financially through the Animal Sponsorship program and assisted with a variety of transport and logistical problems peculiar to a zoo.

Ultimately Quintus became too old to sire cubs and around the same time Taronga Zoo management had decided not to breed any more hybrid tigers and to eventually concentrate on displaying and breeding the purebred Sumatran Tiger. His keepers felt Quintus was also past parade ground duty so he was 'retired' to enjoy the rest of his life free from ceremonial responsibilities. The 5th/7th Battalion still kept in touch from time to time to check on their 'Warrant Officer's' wellbeing.

Sadly, after a very long life, the day came in May 1988 to make that extremely difficult decision to send poor old Quintus to the big parade ground in the sky. Being the daughter of an army colonel, I knew the importance of informing the 5th/7th Battalion of this event. Even though Quintus had ceased being an active part of their battalion life some years before, he was still very much part of that battalion's impressive history.

I telephoned the battalion headquarters and spoke to the Battalion Orderly Room Corporal to explain that we had that morning humanely put down Quintus because of his advanced age and failed health. The corporal asked me what we would do with his body and I explained to him, in great detail, the procedure and routine involved in the complex issue of Zoo body disposal.

Firstly, as with every Zoo animal that dies, there was an autopsy carried out and blood and tissue samples taken for veterinary testing; some organs, body parts and samples would be sent to other scientific institutions for research purposes, the skin would probably be used in the Zoo Education Centre for teaching purposes, the skeleton would be saved for possible display either at the Zoo or the Australian Museum and the remaining body parts would be taken away by our contaminated waste disposal contractor.

After I had finished explaining all of this with sincere compassion as I was personally very sad to see the demise of this old and much-loved tiger, there was deathly silence at the other end of the phone and then the corporal eventually gasped, 'You can't do that, he's a soldier and must have a ceremonial burial!'

We then needed to quickly assemble 'the essence of Quintus' so he could receive this final tribute from his battalion. I asked our veterinarian if she could salvage at least the tiger's heart, whiskers and claws and sent public relations officer Narelle Storey to the Northern Suburbs Crematorium to obtain a suitable urn or container in which we could place the incinerated parts. Narelle was extremely embarrassed when, at the crematorium, in front of grieving relatives who were selecting similar receptacles for their recently departed loved ones, she was forced to admit that hers was for a dearly departed tiger.

The 'essence of Quintus' was subsequently handed over to the 5th/7th Battalion for burial with due army honours. They even created a Quintus Walk and memorial pathway at their Holsworthy headquarters. I always felt very sure that the fighting spirit of this old army mascot was in very safe hands.

FINALLY, ALL THE REMAINING hybrid tigers died out at Taronga and the focus for the past 18 years has been on the magnificent purebred Sumatran species. This is a highly endangered species with currently only 200 in zoos and about 300 to 500 left in the wild. It is the smallest of the tiger sub-species and is deep reddish to dark orange with bold black stripes and cream belly fur. The striped coat provides camouflage for this efficient nocturnal hunter, which rests in the shade or in the water during the heat of the day. The Sumatran Tiger (*Panthera tigris sumatrae*) is found only on the island of Sumatra in Indonesia where its habitat today has been reduced to fragments of forest and swampland.

Taronga-bred Sumatran Tiger cubs have been sent to zoos around Australia and overseas and today Taronga's David Pepper-Edwards, Manager of the Asian Mammals Division, is the Regional Studbook Keeper for this species.

It is always cause for great celebration when an endangered species is born in either Taronga or Western Plains Zoo. It is particularly satisfying to know that the efforts of this State's two Zoos have added to the critically low number of tigers in the world. Tiger cubs are magnetic creatures and Zoo staff, media and visitors are invariably mesmerised by their beauty. And this was never more so than in 1988, when Sumatran Tiger, Meta, gave birth to two healthy cubs.

It wasn't the first time Sumatran Tiger cubs had been born at Taronga. In fact, Meta and her mate, Nico, began producing offspring in 1980. It certainly won't be the last litter but, very sadly however, it will be the litter the Zoo long remembers for the most tragic of reasons.

Carnivore keeper Vicki Scrivener, a senior keeper of nine years' experience and a highly valued member of the Taronga family, lost her life because of her fascination for these cubs.

It was a Saturday afternoon and I remember every detail so clearly. I was unpacking my car having just been food shopping for a dinner party I was planning that evening for friends. I love preparing our house to receive guests and enjoy the shopping, the cooking, setting a special table and this occasion was no exception. It was the lead-up to Christmas, always an enjoyable time of year, and Rob and I were looking forward to seeing two sets of good friends in our home that night.

With my arms full of shopping bags I answered a telephone call at around 2.30 p.m. that changed Zoo life forever. The day's Zoo Duty Officer, Curator Graeme Phipps, was calling to tell me, very briefly and without great detail, that there had been an accident at Taronga and that keeper Vicki Scrivener had been badly injured by a tiger. She was being taken by ambulance to Royal North Shore Hospital at St Leonards. Director John Kelly was on his way to the Zoo and Graeme asked if I could also get there as soon as possible. The bags of shopping were left in a pile on the kitchen floor, and I ran out to the car.

I remember driving to Mosman from my home, about 40 minutes by car from Taronga in Sydney's north, recalling the consequences of when, some years before, a Zoo visitor had climbed a barrier fence to pat a passing tiger. She lost the tips of two fingers and was taken to the Royal North Shore Hospital by ambulance. I learned a valuable lesson then that as soon as an emergency vehicle is involved the media knows about it. Their close monitoring of these radio channels gives them instant information. I prayed I would get to the Zoo before the media wave rolled in.

As soon as I walked into the Director's office, the pale and shocked faces of the gathered staff confirmed my worst fears. We had a very, very serious accident on our hands.

It seemed that, in her lunch hour, Vicki had gone to the Big Cats exhibit to photograph Meta's newborn cubs. She had apparently separated the mother tiger from the cubs and locked her in another den for a brief minute or two while she went in with the cubs to take photos. For reasons known only to Vicki, things went tragically wrong. As recorded later in the Coroner's report, 'a tragic accident caused by the failure of an experienced person to make sure, to the finest detail, that everything was safe before she entered the tigers' den' meant the keeper was attacked from behind by one of the adult tigers.

Several very brave Zoo people were involved in the heroic attempted rescue of their colleague and friend. Vicki was taken to hospital, barely alive, but accompanied by our desperate hopes and prayers for her recovery.

We prepared the most tragic of news statements and John Kelly held the worst news conference a Zoo Director could ever imagine. There were many legal factors to be considered in this scenario so John could only be extremely brief and give no great detail at this time.

I went home about 7.30 p.m. to find my expected dinner guests all in the kitchen cooking their own meal of which I ate nothing. Very early the next morning when our phone rang as I was tidying up the remnants of the previous distorted evening, I hardly dared answer it, as I didn't want to hear the news I expected and so dreaded.

Victoria Robyn Scrivener, aged 33, despite the combined efforts and will of her family, her Zoo colleagues, ambulance officers, hospital and medical staff, had succumbed to her horrendous injuries and died early that Sunday morning. Taronga had lost a highly valued and dedicated member of staff.

Taronga also lost its innocence that most horrific of days. No longer was our Zoo a place of good news and fluffy animal stories. Tigers are top-of-the-line predators and one had put us on the front page of every newspaper in the country and made us the leading news story on every radio and television bulletin that morning. It was the first fatal accident in Taronga's 72-year history.

By the time I arrived at the Zoo on Sunday morning there were about 20 media representatives camped outside the Public Relations office gates. The Zoo Board had decided not to hold a news conference but to only issue a prepared statement. Because the whole tragic episode was now before a court of law the Zoo was unable to elaborate or speculate on any details of the accident or even to publicly reflect on Vicki as a person and a colleague. It was a very difficult time for all Zoo staff, particularly her immediate colleagues, but the pain for Vicki's family was unimaginable.

I really learned who my media friends were then and while I could understand they all had a job to do, I greatly valued the way most of them reported the story. There are always exceptions and one or two journalists made life even more difficult for us all. They pointed out that we always wanted them around for the good news stories but now that the going had got extremely tough we wouldn't co-operate. It was not that we wouldn't cooperate but there was to be a coronial inquiry and many issues and legalities considered. I believe we did the best we could at the time to satisfy the need for news.

Vicki's funeral service was a very public one and her family and friends and her extended Zoo family were left with little space in which to mourn and grieve in private.

The impact of this accident had far-reaching consequences for Taronga and Western Plains Zoos and I believe that some Zoo staff may still struggle with the terrible event that took place that sunny Saturday afternoon in December 1988.

The Board provided all manner of counselling services for staff and the health and safety operations of the Zoo, which were already under review, were overhauled, up-graded and re-written. The public sympathy for Sydney's Zoo was enormous, extremely comforting and greatly appreciated by all the staff.

The Zoo's phone traffic during the days and weeks following the accident was indescribable and so many telephone calls were from members of the community begging us not to destroy the tiger which had killed Vicki. This thought had never even crossed our minds and I think I was quite speechless when I received that first emotional plea. By the time I had answered about 100 such calls I was able to put into the right words the answer that, of course, we would not even contemplate such a fate for the animal. Why would we destroy the beautiful tiger which Vicki loved and which was only behaving like a tiger?

While, understandably, many in the Zoo struggled with their own personal tigers at that time, the Zoo family closed ranks and pulled together in this time of great adversity, showing a united front, especially to the media. I greatly appreciated the strength, energy, wisdom and support I received from my colleagues Narelle Storey, Stanley Edgar, Carol Inkson, Helen Young, Graeme Phipps and John West to enable me to handle that media

SUMATRAN TIGER SPOTS

WHITE MARKINGS ON A TIGER'S ears help cubs to locate their mother in the long grass. Tiger fathers generally play little role in cub rearing and as cubs mature, competition for food and territory means the parents force them to find territories of their own.

pressure in the calmest possible way. My Zoo dreams, even today, invariably involve a tiger.

One of Meta's and Nico's male cubs, Shiva, still lives at Taronga Zoo and he and his mate, the beautiful Seletan originally from Melbourne Zoo, have produced four healthy cubs of their own since 1994. Although they are getting on in years now, these animals are still capable of reproducing and hopes are high that this handsome couple will, at least one more time, add to the number of Sumatran Tigers in the world.

A TIGER OF A DIFFERENT hue arrived at Taronga Zoo in 1992. Chester, a nine-year-old white tiger from Henry Doorly Zoo in Omaha, Nebraska, came to live with us and soon became a huge favourite with visitors and Zoo staff alike.

In the opinion of many zoological purists, white tigers, especially single animals, should not be part of a serious, scientific organisation's collection plan. The space that animal occupies could be better used for housing and displaying purebred Sumatran Tigers, for example, which would be part of a regional species management plan and a worldwide breeding and conservation program.

Zoos must always be mindful of the commercial and educational aspects of their operations, however, and Zoo Director John Kelly, on a visit to the United States, saw the huge potential for Taronga — and other zoos in Australia — if the first-ever white tiger graced our shores. So the handsome Chester found a new harbourside home in Sydney and soon won over even the most strident opponents.

Chester was huge. His weight ranged between 180 and 200 kilograms. His ice blue eyes, beautiful white and black coat, pink nose, enormous feet and graceful tail soon made him an object of great admiration. White tigers are not albino animals, they are mutant colour variations of orange and black tigers. Chester was a white tiger with attitude! He knew he was spectacular. He played to his audience and enjoyed up close and personal interaction with Zoo visitors through the glass viewing window, that's if he wasn't lounging on his bamboo 'divan' bed posing for camera enthusiasts.

He became the 'King of the Zoo' and a very appropriate ambassador for the 'King of the Jungle'. He also became a well-travelled tiger, flying off to spend some time in Western Plains Zoo and also Perth Zoo.

Chester was a great hit during NightZoo at Taronga when he stalked around his jungle domain in the moonlit cool of the evening. He received personal fan mail, birthday and Christmas cards, had sponsors falling over themselves to contribute to his upkeep and the sale of soft toy 'Chesters' sent the retail cash registers into overdrive.

But one little girl from Queensland was his most loyal and lifelong fan. I think she first fell in love with Chester after seeing a photograph and story in the *Australian Women's Weekly*, which heralded his arrival at Taronga. She and her family visited him regularly whenever they were in Sydney and I got to know them quite well. When Chester received anonymous Valentine cards we knew it was Chanel who sent them. She would send him postcards from her family holidays around the world and had her bedroom decorated in Chester memorabilia.

Chester's keeper, Frank McFayden, organised a behind-the-scenes tour of Cats of Asia so Chanel could see her favourite feline close up. Frank even gave Chanel one of Chester's stainless steel milk bowls, which the very large tiger had decided to 'attack' one day. It was ruined beyond use but the huge teeth marks in the bowl meant everything to this devoted young fan.

Unfortunately, by the winter of 1997 Chester's age had caught up with him and he was moved to a smaller enclosure where his keepers could monitor his wellbeing more closely. There were lots of complaints from regular Zoo visitors who lamented that they missed seeing the old cat on his usual 'throne'. Chester had several health problems during his long life and was often a regular on veterinarian Larry Vogelnest's daily rounds. But Chester always seemed to bounce back, testimony to the tender loving veterinary and keeping care he received.

Ultimately, at the grand age of 18 years, dear old Chester had to be euthanased and it was a very difficult phone call I needed to make to Chanel and her family. Chester's passing left a huge hole in the Zoo and everyone missed this special cat, but perhaps none more so than his little Valentine from Brisbane.

Handsome Kibabu, Taronga's Western Lowland Gorilla silverback, came to Sydney from The Netherlands in December 1996.

Mother's Day chrysanthemums receive careful scrutiny from Western Lowland Gorilla female Kijivu, before being tasted and then spat out in disgust.

With the bunch of chrysanthemums as a prop, Kijivu shows off her natural balance and poise.

Mary, a Mueller's Gibbon, is 'queen of the Zoo'. She reigns from high up in a large Moreton Bay fig tree. Her early morning hoots herald the beginning of each new day at Taronga.

'Outdoor furniture' in Taronga's Chimpanzee Park, such as ironbark trees and strong ropes, needs to be renewed approximately every five years because of wear and tear from the boisterous chimps. Lubutu, an 18-month-old chimp, gives some new ropes the swing test.

strong leather glove for their veterinary checks, even at six weeks of age.

Taronga's Meerkats enjoy basking in the warmth of an infrared heat lamp, which has been installed in their Zoo 'desert'. I think the Meerkat in the centre of the photograph may have just set his little tail on fire!

Fly, be free! In December 1999, a South Pacific cruise on board P&O *Fair Princess* enabled Taronga's Wildlife Supervisor Elizabeth Hall to return a rehabilitated Tahiti Petrel to its home range near New Caledonia.
(Photograph courtesy P&O Cruises)

A Wandering Albatross, somewhat reluctant to return to life on the ocean waves, is released from the cliff tops of Sydney's North Head after recuperating in Taronga Zoo's wildlife clinic.

A resident of Taronga's Koala Walkabout does not need any cute lessons.

Two Taronga Giraffes play with raindrops on a wet Sydney day.

CHAPTER 5

Dances with Brolgas

BIRDS EPITOMISE FOR me my wonder of nature. The incredible number of different species, colours, their adaptations, habits and behaviour, birdsong and the absolute mystery, to me, of flight, never fail to inspire and delight.

Even as I write this book I find myself distracted constantly by movement in my garden as Rainbow Lorikeets, Eastern Rosellas, Magpies, Eastern Spinebills, Noisy Miners and Butcher Birds come to swim in or drink from the bird bath under the trees just outside my window. Hot days or cold, sunny days or in rain, they are out there going about their busy birdie business.

They raise a myriad of questions in my mind as I wonder about their eyesight, which is so keen it enables them to see a tiny grub on the ground as they perch high in a tree. Then there is the often rainbow hues of their plumage that must surely mean they see colour like we do. And how does that tiny throat produce notes and musical sounds a world-class opera singer or an accomplished musician can only emulate? And what do all those different sounds and calls mean, and are they sometimes done just for the bird's own pleasure?

Australia is so fortunate to have more than 670 different species of birds that breed in or regularly visit this country, and Taronga Zoo has long had a commitment to displaying as wide a variety of those species as possible as well as birds from other lands.

Taronga has also had a long history of exceptional bird specialists whose aviculture and ornithological knowledge have been envied around the world and their commitment to their feathered charges absolute. Their desire to share with Zoo visitors this passion for birds has ensured that Taronga's aviaries have always been little centres of excellence in zoo keeping and display.

Les Clayton was the 'Bird Man of Taronga' for over 40 years and had been at the Zoo for about 25 of those when I arrived in 1975. I always thought of him as our 'national treasure' and a Zoo 'tribal elder'. Everyone loved Les of the twinkling blue eyes and cheeky smile. He was a gentle and quiet man, who not only toiled long and hard for the benefit of Taronga's bird collection but also found time to answer pleas for help from all over Sydney to rescue native birds in distress any time of the day or night.

I got to know and rely on Les right from the beginning of my Zoo days. It was Les I would turn to for advice on the many phone calls from members of the public that I handled through Taronga's Animal Enquiries service. This daily telephone service, now operated by Zoo Friends Volunteers, gives the community an opportunity to contact Taronga and receive advice on countless numbers of wildlife problems and issues ranging from animal identification and the care of orphaned native animals to the removal of pesky possums in the roof or, alternatively, how to attract them and other wildlife to your garden.

Being the dispenser of this twice-daily dose of vital information was a steep and sudden learning curve for me as I began my Zoo career from a zero animal knowledge base. Although I relied heavily on the 'Zoo Bible' — a loose-leaf folder of helpful animal hints compiled and added to over the years perhaps by former struggling public relations officers — I soon learned that I could rely on dear Les Clayton for all manner of answers to birdie questions. He was always available and nothing was ever a trouble.

Initially, along with my Animal Enquiries duties, I also had to carry out the Zoo's Animal Reception tasks and was the repository for the daily stream of sick, injured or orphaned native animals that members of the public brought to Taronga for care. These animals were usually victims of cat or dog attacks, road accidents, or other miscellaneous urban perils.

Many birds were accident victims and found their way to my office. It was Les who came to collect them. His comforting hands would soothe the injured bird as he helped me, in my early days, to identify each species and fill out the dreaded arrival form before the patient was put in a box or, more often than not, tucked safely inside Les's khaki shirt and taken off to the Zoo Veterinary Hospital or Les's 'outpatients ward' in the Bird House for assessment and treatment.

I used to look at Les and wonder how anyone could appear so dishevelled so early each morning. His uniform looked like he had slept in it and he always looked like he was covered with birdseed and bird pooh. It wasn't until I visited the Zoo Bird House and saw Les's 'office' that I realised that all of the above was very true. Les often slept the night at the Zoo surrounded by boxes, cages, small aviaries and humidicribs full of newly hatched chicks that needed his intensive care to survive. His bed was a tiny camp stretcher wedged under rows of bird boxes. A restless night by seed-eating birds meant Les started the day covered with seed from head to toe.

Much of Taronga Zoo's world-recognised reputation for successful bird breeding was as a result of Les's dedication and expertise. Les also enjoyed success in his own breeding program as he and his wife, Pat, had seven children but goodness knows how, as Les always spent at least half the year, especially during late winter and spring, at the Zoo hand-raising his feathered 'children'.

The successful hatching of native and exotic birds at Taronga was always a great delight to Les and provided the Zoo with wonderful media opportunities. Photographers and television crews would always flock to Taronga to meet the newly hatched kiwi, brolga or macaw. Les was always so proud of his chicks and the media loved him for his enthusiastic co-operation with their every request at these publicity calls.

ONE PARTICULAR FAVOURITE of Les's was the Blue and Yellow Macaw (*Ara ararauna*) chick that hatched in an incubator in December 1983 and was hand-raised by the doting Les. This bird, which Les named Eleni, was the first Blue and Yellow Macaw ever bred at Taronga Zoo. Macaws are large, brightly coloured parrots from the rainforests of South America and are an endangered species, so this bird was very precious indeed. Apart from the destruction of their habitat, the adult macaw is killed for its feathers and the chicks are often taken for the live bird trade. They attract attention by their vivid colouration and spectacularly strident call. Macaws are large parrots measuring about 90 centimetres and weighing about one kilogram.

Les spoiled Eleni from day one, feeding her every few hours and only giving her the very best tropical fruit and nuts all mashed up and fed from a specially fashioned silver spoon made to resemble her mother's beak.

I have to say I never really warmed to Eleni and I don't think she was too fond of me, either. In fact, I think she went out of her way to cause me grief. As I reflect on my Zoo memories the ones involving Eleni invariably involve angst.

Initially we got on quite well as she was perfectly behaved when she made her media debut as a rather ugly hatchling with a 'five o'clock shadow' of pinfeathers, a huge beak and an extremely loud shriek.

Some months later, the Channel 9 'Midday Show' asked if Eleni and Les could appear live on the program and I saw this as a great opportunity to promote the upcoming Zoo Month. By this time, Eleni was fully coloured, looking resplendent in her blue and yellow flight feathers — and she knew it. We drove in the bird truck to the television station at Willoughby and I don't know how I made it there with my right ear still intact. Les insisted on Eleni travelling freely in the cabin with us and not in a cage in the back like most other birds. She obviously didn't like me being in the truck with Les or was, perhaps, jealous of my gold earrings as she spent the whole trip either shrieking her disapproval or using her can-opener beak to try and prise the gold studs from my ear lobes.

We finally arrived at the studio and Les and I headed straight to make-up. Les was in and out of the make-up room in a flash but it took a good deal longer for the Channel 9 make-up artists to conceal my shredded ear and make me look presentable. Les had great delight in telling everyone later that it only took three minutes for he and Eleni to be made up for their television appearance while it took 30 minutes for me to receive the necessary repair work! I think the make-up girls worked their magic in double-quick time on Les just to get his shrieking macaw out of earshot and away from their rouge pots.

To my amazement, Eleni was an angel right through the live interview with wonderful host Brian Bury. She was a perfect, and very quiet, Zoo ambassador allowing Les and I to talk about the history of Taronga, the Zoo's upcoming birthday and Zoo Month celebrations.

I should have known it was too good to be true. At the end of the segment just before the show broke for a commercial, Brian asked the audience to thank Les, Eleni and I, which they did with enthusiastic applause. Les must have been lulled into a false sense of security and had relaxed his grip on Eleni's feet. She got a fright at the loud clapping and took off to the ceiling of the studio just as an ad for barbecued chicken went to air! I could just see it, this precious bird being 'cooked' by the hot studio lights.

We were pushed off the set by the floor manager as he prepared for the next guests and as the audience offered all sorts of helpful advice, Les and I had approximately three minutes to encourage the startled parrot back down from the rafters. Les had a pocket full of peanuts in the shell and as soon as he put one in his mouth and Eleni saw it from a great height she flew down to her beloved Les. He got a firm grip on her feet and we were marched out of the studio and back to the truck for the return journey to Mosman. I couldn't help thinking that Eleni looked very pleased with the impact of her first live television appearance. I also seem to recall that the 'bar' in the Zoo Public Relations office opened rather early that evening.

Eleni was the cause of a visit to the doctor for me in December 1985. She was now almost an adult and was living in an off-exhibit aviary awaiting a suitable partner to be found for her for future breeding purposes. She must have got tired of waiting and decided, one hot summer night, to chew her way out of the aviary and fly off in search of a partner herself. Imagine poor Les's dismay when he found his precious bird missing in the morning.

The search was on. Zoo staff spent the morning combing the grounds, peering at treetops hoping to find Eleni who, it was thought, would not have flown far. A day and a night went by but there was no sign of the macaw.

It was the weekend before Christmas and, as usual, I had left all my gift and food shopping until the last minute and had planned to complete it over those two days. Unfortunately, Eleni's escape put paid to that as we went public about the missing bird in the hope that the community could help us find her. We held a news conference at the Zoo on the Friday morning and I set up the 'Macaw Hot Line' on which we received calls, day and night, from all around Sydney with possible, and improbable, sightings of the colourful parrot. The Zoo's hardworking Zoo Friends Volunteers took the calls hour after hour, but all to no avail. The weekend went by in a blur of phone calls and media enquiries and not a macaw or a Christmas present in sight. By about day five, the day before Christmas Eve, Les had just about given up hope when out of the sunset came The American Eagles to the rescue.

The American Eagles High Dive Team was performing at Taronga for the summer holidays. They were an incredibly talented and entertaining group of high divers from the United States who plunged from a 25-metre high tower into the three-metre deep seal pool three times a day. After the 6.00 p.m. performance, team leader Chip Humphrey reported seeing, from his very high perch, a large colourful bird in the treetops near Taronga's rainforest aviary.

FEATHERED FACTS

♠ Total number of birds at Taronga Zoo: 1,068
♠ Number of species/sub-species: 167
♠ Largest bird at the Zoo: the Emu at 40 kg
♠ Smallest bird at the Zoo: The Variegated Fairy Wren at approximately 9 g
♠ Most vocal bird at the Zoo: the South American Sun Conure (unpleasantly noisy but it makes up for it in beauty!)
♠ Number of bird meals prepared each day by Taronga bird keepers: 200
♠ Fussiest eater in the Taronga bird collection: the Glossy Black Cockatoo, which in the wild eats primarily casuarina nuts, which are difficult to obtain

Les raced to the reported vicinity with the obligatory peanuts, and began calling Eleni's name. After about five minutes a very bedraggled, thin and rather shame-faced looking macaw lurched down out of the trees and landed on Les's shoulder. There were tears all around as the odd couple were reunited. Eleni had green stains around her beak and on her black-and-white face indicating she had resorted to eating such mundane things as leaves and bark in place of her usual fare of roasted peanuts and tropical fruit.

I went straight back to the office and put out a news release announcing the successful outcome and early the next day Les and the dreaded Eleni were again the centre of media attention at a news conference as the happy ending made a great Christmas story.

I finally did my entire shopping sweep for presents and food on Christmas Eve, which I followed up with a visit to the doctor to see why my heart felt like it was leaping out of my chest. He confirmed that I was suffering from stress-related hypertension which he sympathetically said was quite common around Christmas. I couldn't help thinking mine was more likely parrot-related.

I should add that a suitable mate was found for Eleni and she has subsequently increased the world's Blue and Yellow Macaw population by five, which is great news for the species. She has also been the model of a perfect mother and the exploits of her rather colourful youth are now only a distant memory. They are still, however, firmly fixed in mine.

IF I TELL ANOTHER story about a lost bird and Les Clayton, it might sound as though Les was careless with his charges. In fact, the exact opposite is true. He went to extreme lengths to ensure that the birds in his care received only the very best of everything. It was just that he had so many birds over so many years and all of them had wings, and as wings are used for flying, sometimes that meant accidentally flying away from Les.

Bluey brolga was another of Les's hand-raised birds, and as far as this bird was concerned Les was its mother. As a chick, Bluey would follow Les all around the Bird House not letting 'mum' out of her sight for a minute.

Brolgas (*Grus rubicundus*) would have to be one of my favourite birds. I love their long, delicate legs, their pointy chopstick-like beak, their impossibly beautiful slate grey plumage and contrasting red head and neck. Brolgas are most abundant throughout coastal tropical Australia, close to swamps. They are sometimes known as the Australian Crane and the Native Companion. Males grow to about 1,250 millimetres in length and the female to 1,150 millimetres. The appearance of both sexes is similar.

The brolga is so stately and graceful. I love their elaborate display routines and the fact that they seem to dance and trumpet for the sheer joy of it. I once performed in a ballet at the Sydney Conservatorium of Music that was based on the Aboriginal legend about a young girl who is turned into a brolga and dances for the rest of her life. Perhaps that is why the Zoo brolgas crept so easily into my heart.

Bluey was an orphaned brolga who was found at Evans Head in northern New South Wales in December 1972 and brought to Taronga Zoo. Under the expert care of Les and his team of bird keepers, Bluey grew into a very handsome adult bird. She eventually took up residence in the wetlands exhibit, just inside Taronga's main entrance. Her flight feathers were clipped from time to time to prevent her flying too far, but for some reason it had been a while between clips and Bluey's feathers had grown somewhat. So when a Zoo workman started up a jackhammer early one morning, Bluey took fright and flew off and was last seen heading west in the direction of the Harbour Bridge.

Just the mere thought of a 'Brolga Hot Line' to record any possible sightings of Bluey filled me with dread but, thankfully, Les decided to wait a little while before seeking community help. He was convinced Bluey would not fly far as she had no flight practice and her muscle strength would be lacking. Sure enough, about 24 hours later, I received a phone call from a rather excited local Mosman resident. She had been walking her dog on Whiting Beach — the sandy strip in a pretty little bay just west of the Zoo boundary wall — and came upon what she described as a 'large, grey, friendly flamingo'. I knew immediately that it had to be our Bluey. I drove to the Bird House, picked up Les in my little yellow Mini-Minor Zoo wagon and hurried to Whiting Beach.

Sure enough, there was Bluey standing at the other end of the beach looking very bewildered and lonely. Then she saw her precious Les and ran the full length of the beach at top speed on those long spindly legs with her wings outstretched to the side. As she reached us she wrapped her wings around Les's entire body and quivered with relief and joy at finding her lost 'mum'. We bundled Bluey into the back of the minivan and I cried all the way back to the Zoo. I think Les was doing a bit of snivelling, too. I should note here that Bluey the brolga eventually forgot her 'fixation' for Les and paired up with Paddy the brolga to produce a chick in November 1981, subsequently named Scrapper the brolga.

Les Clayton retired from Taronga Zoo in February 1993 after 48 years of outstanding service and sadly passed away just three years later. I am glad to say the unique Clayton spirit lives on at Taronga in the form of Les's equally kind and generous son, Lindsay, who works his own special magic maintaining the Zoo exhibits.

NOT ALL TARONGA'S birds are Zoo-hatched. Some are donated, as was the case with Henry the Gang-gang Cockatoo who arrived at his new home at Taronga in October 1986. Henry's arrival card notes 'He was found 32 years ago at Batemans Bay as a juvenile. He had been shot in the right wing and he can't fly.'

COCKATOO FACTS

THE GANG-GANG (*Callocephalon fimbriatum*) is also known as the Red-crowned Cockatoo and Helmeted Cockatoo and has beautiful slate-grey plumage on the wings, breast and tail edged with white. Only the male has a bright red head and crest. They are found from the seaboard to high inland mountain ranges of South East Australia throughout heavily timbered and wooded eucalypt areas where they feed on seeds and berries. They grow to about 350 mm in length and their call has been described as 'a prolonged jarring croak'.

Henry the Gang-gang had been a lifelong companion pet and had outlived his owner. The elderly, deceased gentleman's family found a new home for Henry at the Zoo. As the bird was well-used to being handled, it was decided that he would be very suitable for use in the classrooms in Taronga's Education Centre where visiting school children — up to 25,000 each year — have lessons on a wide variety of exciting and stimulating animal-related subjects.

Henry's first classroom appearance was almost his last. As Steve McAuley, the Education Centre Manager, brought him into the room to introduce Henry to a group of young students from a Catholic girls convent school, the bird began using a stream of language that would have made the proverbial sailor blush. Henry was immediately expelled and kept out of the classroom until his bad language had been 'unlearned'. He has subsequently been on his best behaviour for many years and is a star of the Zoomobile, Taronga's outreach program. This educational travelling 'zoo' and the Zoomobile officer go to a variety of schools whose students, for one reason or another, cannot visit Taronga. Henry particularly likes visiting prisons and reform schools.

WESTERN PLAINS ZOO'S network of lakes, streams and moats provides a haven for hundreds of free-ranging birds. The zoo is also the home of a very successful conservation program for the threatened malleefowl.

The malleefowl doesn't realise it but it has another name, occasioned by a royal 'flush'. Queen Elizabeth and Prince Philip visited Western Plains Zoo in February 1991 and, as is quite often the case with royal visits to this country, the mercury reached almost boiling point that day. Thousands of local school children were assembled in the central western high-summer sun to greet the royal party. The Mayor of Dubbo and local councillors had joined other assorted dignitaries and the school kids who had travelled far to join Zoo Board Members and staff in welcoming Her Majesty and Prince Philip to this country's premier open-range zoo. It was an exciting occasion and we were proud to be part of the tour itinerary.

The Chairman of the Board, the very genuine and proudly enthusiastic Bruce Robertson, was to give the welcoming address and wanted to use this unique opportunity to great effect by detailing the exceptional efforts of Western Plains Zoo in the areas of conservation of endangered species, particularly for Australian native animals.

The list of recovery and conservation programs for native animals carried out at Western Plains Zoo is long and complex and includes work with the Greater Bilby, Mala, Eastern Barred Bandicoot, Northern Hairy-nosed Wombat, Bridled Nailtail Wallaby and last, but most successful of all, the program for the quiet and unobtrusive malleefowl — a bird that is known to do more flapping than flying.

By the time the Zoo Chairman came to the mention of the Malleefowl the enormity of the occasion, coupled with the 42 degree heat caused the name malleefowl to completely disappear from Bruce's memory, and the royals seemed singularly unimpressed with the fact that Western Plains Zoo 'had a very successful conservation program for the highly endangered hen hen'! The royal dais shook as Zoo staff tried to stifle their giggles, and from then on the malleefowl at Western Plains Zoo have been fondly known as the hen hens.

THE ZOO'S MALE ANDEAN Condor is also a bird with an alias. He was hatched at Taronga in September 1979 and hand-raised by Bird Curator John DeJose. When he made his media debut not long after hatching he was called Diablo, a fine Spanish name for a bird that was proving to be a bit of a devil to feed every two hours with finely mashed day-old chicks and blended-up mice.

Taronga Zoo has displayed Andean Condors since about 1935. When newly hatched, Diablo looked like a powder puff with a beak. This huge ball of fluffy whitish-grey down was the centre of much attention as he was the first Andean Condor ever hatched at Taronga.

Malleefowl Facts

THE MALLEEFOWL (*Leipoa ocellata*) originates in the dry inland scrub of southern Australia where their numbers have been diminishing substantially because of habitat destruction and from predation by introduced and feral species. It is an interesting bird because it mates for life, and the male malleefowl must maintain the nest mound year-round while the female wanders off in search of food, which will sustain her production of eggs. Each pair of birds occupies a permanent territory of 40 to 70 hectares of mallee scrub and in it may have up to five scattered nest mounds, only one of which is used in a season.

At the beginning of the nesting season in autumn, both birds build a new mound on the site of an old one, which they scrape out to a depth of about one metre and a diameter of 3 to 4 m. They collect and scrape leaf litter, bark and twigs into this hole and then cover this with dirt kicked back with their powerful feet. The mound also contains an egg chamber and after rain has wet this area, it is covered with sand. All this work takes about four to five months of intermittent activity before the mound is ready to receive the eggs, the number of which is dependent on rainfall. Throughout the incubation period the male maintains the nest mound and defends the area around it. He must also keep the fermenting material at a constant temperature of 33°C. Ultimately the heat from the fermentation process hatches the eggs and the young are independent and free running at hatching.

The malleefowl conservation program, run in conjunction with scientists and biologists at the New South Wales National Parks and Wildlife Service has, since 1988, been one of the flagships of Western Plains Zoo's efforts to save numerous species of endangered native and exotic animals. About 1,000 malleefowl chicks, either hatched at the Zoo from eggs gathered in the wild or bred at the Zoo, have now been released into sanctuaries and reserves in western New South Wales. The numbers have built up considerably thanks to this conservation program initiated by, among others, innovative and tireless Zoo Curator Graeme Phipps, Bird Keeper Ken Delamotte and progressed now by the Conservation Biologist at Western Plains Zoo, Phil Cameron.

He was an instant media star and was invited to appear on the Channel 9 'Midday Show'. I felt confident that Diablo would be safe and not end up in the rafters and overhead lights, as unlike Eleni the macaw he did not yet have his flight feathers. He did, however, exhibit some form of weird shoe fetish on camera. He jumped down from John DeJose's knee and attacked host Brian Bury's best alligator-skin loafers, shredding them with his powerful condor beak. Maybe he recognised a former inhabitant of the wilds of South America?

Diablo soon had a name change and became Bruce. Zookeepers do things like that for some odd reason. For the past 22 years Bruce the Andean Condor has been a well-known Taronga identity. He was the inaugural star of the Free Flight Bird Show and it was amazing to see this huge bird with a three-metre

CONDOR FACTS

ANDEAN CONDORS (*Vulture gryphus*) are the largest flying birds of prey in the world. They live in the Andes Mountains of South America and can soar to altitudes of 6,000 m. The adult male can weigh up to 12 kg and have a wingspan of over three metres. Fully grown, the Andean Condor is glossy black with white patches on the wings and a ruff of fine snowy white feathers around the neck. The male has a reddish comb and wattle on the head and neck. They are mainly scavengers and, despite their great size and weight, cannot kill anything but a weak or dying animal.

Condors mate for life. A single egg is laid which both the male and female incubate. It then takes two years to successfully rear a young condor. The Andean Condor was once listed as Vulnerable but action has been taken to conserve the bird's habitat and preferred breeding sites.

wingspan do his 'B52 bomber' impersonation over the heads of the enthralled audience. But it eventually became just too difficult to carry Bruce in a giant pet pack twice a day from his aviary to the show site for take-off. Now, however, he and partner Connie the condor, hatched at Taronga a few years after Bruce, have been given an expansive new aviary adjacent to the beautiful amphitheatre overlooking the harbour and Bruce, much to everyone's delight, is back in training for the Free Flight Bird Show. I was overjoyed to learn recently that Bruce and Connie laid an egg shortly after moving into their palatial new home. In January 2002 this egg was successfully hatched in a Zoo incubator and Taronga has another 'powder puff with a beak'.

THE MORE RECENT and innovative developments in the way Taronga displays the huge diversity of birds now enables Zoo visitors to appreciate the magic and mystery of flight even more.

The Free Flight Bird Show was launched in September 1997 and the new band of bird keepers like manager Kevin Evans, team leader Matthew Kettle, and his group of dedicated and talented presenters, train an amazing variety of birds from White Breasted Sea Eagles to magpies, owls, galahs and brolgas to demonstrate, on cue, each bird's unique adaptations and behaviours. I believe, as a result of this wonderful, informative and entertaining show, that birds are receiving the sort of recognition they deserve.

The 1,000-seat amphitheatre — the unique venue for the Free Flight Bird Show — is built on the side of the Taronga hill, looking over beautiful Sydney Harbour, with the impressive cityscape as the backdrop. The venue is breathtaking but, once

BIRD DEPARTMENT'S WEEKLY FOOD SHOPPING LIST

21 KG APPLES

7 KG BANANAS

3 BUNCHES BEETROOT

1.4 KG BROCCOLI

1.4 KG CARROTS

1.2 KG CELERY

14 CORN ON THE COB

300 EGGS

42 BUNCHES ENDIVE

3.5 KG KALE

7 KG KIWI FRUIT

10.5 KG ORANGES

25 KG PAWPAW

7 KG PEARS

500 G PEAS

1.75 KG SWEET POTATO

7 KG ROO MINCE

2.1 KG HEART

7 KG WHITEBAIT

40 KG RED SPOT WHITING

VAST QUANTITIES OF SUNFLOWER SEED, PARROT MIX, WATERBIRD MIX, PIGEON MIX AND FINCH MIX

7 LOAVES BREAD

Simplest way to attract birds to your garden: install a bird bath and keep the water fresh by cleaning it out every couple of days.

the show begins, you soon find you are totally absorbed in the birds. Barking Owls whistle overhead, the Sea Eagle swoops, the Black Breasted Buzzard opens emu eggs with stones, Bruce the Andean Condor almost blocks the sun with his enormous wingspan and colourful Australian parrots brighten the skies with their jewelled plumage.

I even find myself applauding the Blue and Gold Macaw, despite the fact that it is an offspring of Eleni! All the while, the presenters are using every opportunity to pass on fascinating facts about the world's bird life. Zoo visitors of all ages and from all corners of the globe love this show and it has quickly become a highlight of a Taronga visit.

Billy the brolga is one of the stars of the show and also one of my very special Zoo 'friends'. He and his keeper, Cathy Alexander, would regularly visit my office as part of Billy's training routine. Walking Billy around the Zoo grounds helped him become familiar with different people and unusual situations. Billy would carefully walk up the stairs to my first floor office and potter around the room closely inspecting everything while Cathy and I chatted. The entire contents of my desk, my keyboard, coffee table and bookshelves were regularly and carefully inspected by this inquisitive bird and he often provided a most unusual welcome to visitors to the Zoo Public Relations office. When the local papers and Channel 9 'Today' wanted to do some farewell photos of me with a Zoo animal during my last week at Taronga I quickly nominated Billy the brolga as my 'dancing partner'. These photographs and my memories of beautiful brolgas will be lifelong treasures.

CHAPTER 6

Why Do You Love Koalas?

MANY AUSTRALIANS often take their unique wildlife for granted. We are all somewhat guilty of being a little blasé about the animals with which we share this huge continent. Perhaps it's because many of this country's 'critters' are nocturnal and are out and about when most of us are asleep. Many animals are, therefore, quite unfamiliar.

The beautiful birds we see around us every day are easy to appreciate and because Australia has the largest number of dangerous and venomous flora and fauna in the world the 'shock horror' creatures like sharks, box jellyfish, stonefish, crocodiles, venomous snakes and spiders that attain infamy through a high media profile are well known to us all. It's mainly the small brownish-grey marsupial mammals that hop, scurry, glide and climb around under the cover of darkness that often remain a mystery but which help make Australia the unique country it is.

Koalas are high on the must-see list for many overseas visitors to Taronga, probably mostly thanks to this country's

Koala Facts

KOALAS LIVE IN EUCALYPT forests along the eastern seaboard of Australia. While the koala (*Phascolarctos cinereus*), which is a marsupial, is often incorrectly called a 'Koala Bear' it is not related to bears, which are placental mammals. A koala's average lifespan is around 12 years but they have been known to live up to 15 years in captivity. They are solitary animals with each animal having a territory of about .5 to 1.5 ha, depending on the availability of food trees.

Koalas breed only once a year with mating occurring during spring and summer. At this time, the males call very loudly, heralding their presence. A single young (very occasionally twins) is born after a gestation period of 35 days and crawls its way to the mother's pouch where the hairless 'joey' develops for the next six months. After that, it rides around on its mother's back, returning to the pouch to feed. After 12 months it is permanently out of the pouch and by the time it is two years old it is completely independent and takes off to seek a territory of its own.

Koalas enjoy a very laid-back image. The koala sleeps in a tree fork for most of the day and moves around at night to feed for about four hours. They are their most active just after sunset. The koala sleeps for up to 20 hours a day, largely because their diet is so energy poor. They are great energy conservers.

international airline's advertising campaign in the United States. Unfortunately some people do get the koala's name a little confused as I discovered when an American Zoo visitor asked me one morning if I could point her in the direction of 'those cute little Qantases'!

In the mid–1970s I saw the first koala born at Taronga for about ten years. This fuzzy grey joey with its black 'rubber' nose, soft white chin and stomach, and tiny brown eyes, was the centre of media attention right from the moment she began peering shyly out of her mother's pouch. This joey was named Amelia after Mrs Amelia Green, an elderly woman who lived on a small land holding in the outer Sydney suburb of Kenthurst. Mrs Green very generously offered access to the zookeepers to collect eucalyptus browse for the Taronga's koala colony.

Every day the Zoo koalas munch their way through the tips of three fresh eucalypt branches each and the Head Keeper of Australian Mammals Dave Thomas and his team were kept very busy collecting enough branches to maintain supply for the very fussy eaters. Koalas live in and eat almost entirely eucalypt trees. In other words, they can literally eat themselves out of house and home. They have been known to feed on some non-eucalypt species such as black wattle but this is most likely because nothing else is available. Koalas seldom drink as they generally obtain enough water from the leaves.

Eucalypt trees around the Taronga Zoo grounds and car parks were often used as a food source as were those in local council parks and gardens. Sometimes we made appeals via the local media for additional supplies and residents always responded by offering the appropriate trees in their gardens as collection options.

The more koalas the Zoo has, the more leaves are needed. As the koala colony numbers grew local schools also became involved in planting trees which ultimately could be used as a food source for the koalas. About five different species of koala-preferred eucalypt trees were propagated and provided by Taronga and planted and cared for by school children in playgrounds from Mosman to Mona Vale. It was a very practical way to encourage their interest in Australian native animals and provided Taronga's koala keepers with a fresh and convenient supply of leaves for their animals.

MOST PEOPLE ONLY ever see a koala in a zoo or wildlife park and that view is one of a very relaxed, benign animal that spends its life either munching on eucalypt leaves or sleeping in the sun. Other images are formed through childhood storybooks and toys that invariably ensure an image of a koala as a soft, cuddly and friendly creature.

Koalas, in reality, have extremely sharp, razor-like claws designed for clinging to trees, and strong grinding teeth for gnawing through tough leaves. They can be noisy, feisty and difficult to handle and when Amelia the koala was still only a youngster I saw her inflict quite a severe injury on a Taronga Zoo keeper, Lee Moyes.

The joey was separated briefly from her mum for her first veterinary checkup and was not a happy koala. All she wanted to do was get back to her mother and cling onto something familiar and secure. The necessary ordeal over, the keeper was transferring her from a carry sack back to a tree in the Koala Walkabout. She did an almighty, but greatly misjudged, leap and landed half on the poor man's face instead of the tree trunk. The result was a keeper with deep razor-like cuts to his face. I immediately saw koalas in a new light. They are certainly not the cuddly toy I once thought they were. They are definitely cute but dangerous, too.

THE KOALA WALKABOUT exhibit at Taronga puts people up in the trees with the koalas and provides wonderful viewing opportunities. It is always a popular stopping-off point with local and overseas visitors. The Koala Walkabout was one of the first new exhibits created at Taronga Zoo by Zoo Director Ronald Strahan in the late 1960s and early 1970s. This simple and elegant Zoo exhibit is still a showpiece today. Zoo archives also record that it was the first of the sponsored structures at Taronga and that the Koala Motel chain provided $20,000 towards its cost.

People walk up an elevated ramp to 'ooh' and 'ahh' at the curled-up balls of grey fur. If a sleepy koala so much as blinks there is a huge wave of activity around the viewing walkway as visitors come rushing to see what all the excitement is about.

Over my 25 years at the Zoo I spent a lot of time in this particular part of Taronga. It was a popular location for media interviews on a range of subjects. It typified Taronga and we could provide access to the exhibit and have animals in near proximity while still being reasonably safe. Two incidents remain vividly in my mind although they are the kind of 'memories from the Zoo' I believe I did try to repress.

The first one involved a hand-raised Red-necked Pademelon, which is a small, greyish fawn-coloured macropod that is found quite commonly in the dense rainforests and eucalypt forests of eastern Australia. This particular Red-necked Pademelon's mother had been killed by a car and the orphaned joey had been brought to Taronga for hand-raising. It was now weaned and was spending its days in the Koala Walkabout exhibit waiting just a little bit longer until it was big enough to join the Zoo's Red-necked Pademelons in their exhibit. It usually hopped around the shaded floor of the exhibit, innocently minding its own business under the lush tree ferns while the koalas lived in the treetops above.

It was early 1984 and the Deputy Director of Taronga Zoo, Dr David Butcher, was in the koala exhibit with a Japanese television crew doing an interview on the strong possibility in the near future of koalas being sent from Sydney's Zoo to Tokyo and Nagoya zoos. As you could imagine, a certain amount of hype was beginning to build around this animal transfer and, quite naturally, this interest was being led from the Japanese end of the equation.

David Butcher, an enthusiastic, knowledgeable and articulate Zoo spokesman, was about a third of the way into the rather searching live interview on the animal welfare issues involved in transporting the koalas when the resident Red-necked Pademelon which was, until that point, being totally ignored, decided that David's fawn-coloured suede desert boots looked quite attractive. The rather feverish-looking little pademelon shot out from the ferns and proceeded to jump onto David's left foot and 'mate' it extremely vigorously. Just as that happened, the television director ordered a wide shot and David was seen all over Japan, spluttering something about international goodwill and koalas being 'ambassadors for conservation' while trying unsuccessfully to shake off a very amorous Red-necked Pademelon which was 'married' to his boot.

THERE MUST BE SOMETHING about the Koala Walkabout exhibit that encourages embarrassing scenes because I remember another such incident that involved mating koalas.

The best time of day to view the koalas to guarantee at least some semblance of activity from these diurnally inactive animals is late in the afternoon. I always suggested to the media that they photograph or film there around 3.30 p.m. when the koalas received fresh leaves and would be moving about searching out the juicy new eucalypt tips. One particular spring afternoon we got more 'activity' than we bargained for and I heard one of the most bizarre sex education lessons ever devised.

We were all set up with the television camera trained on a particularly pretty little female koala that was feeding contentedly in a tree fork. There were also quite a few late afternoon visitors at the Koala Walkabout enjoying what is a very pleasant time of day in the Zoo. The next thing we knew, the male koala a tree or two away started calling loudly and must have decided that our photogenic female koala looked very pretty, too. He leapt from his tree, landed just behind the unsuspecting female and began grunting and growling his way onto her back. She looked shocked and horrified and tried in vain to escape but the male had a good grasp and was determined to succeed in his mission.

The nature documentary television crew was overjoyed at the unique footage they were obtaining, but the father of one very interested seven-year-old bystander had to do some fast thinking when his son asked, 'What are the koalas doing, Daddy?' The extremely inventive explanation, 'The koala in the front is blind and the one behind is helping her home', is indelibly recorded in my Zoo diary!

IN 1983 THE NEW SOUTH WALES Government working in conjunction with the Australian National Parks and Wildlife Service gave approval to export zoo-bred koalas to selected

Japanese zoos, and Taronga and Western Plains Zoos began a long and rewarding relationship with two of Japan's finest zoological parks. New South Wales enjoys a sister-state relationship with Tokyo and Sydney is the sister-city of Nagoya so their zoos were selected to receive the first koalas to be exported from this State since the late 1950s.

The preparations required for such a unique animal transaction and transfer are understandably complex, often difficult and tied up in kilometres of red tape which could stretch all the way from Australia to Japan. It took a huge amount of international understanding, diplomatic negotiation, determination, good humour, patience and goodwill on the part of both countries to eventually find our way around many obstacles, some of which were as tall as a eucalypt tree and as thick as a forest. But all good things come to those who wait and pray and plant eucalypt trees.

Ultimately all the rules and regulations for the export were devised, the stringent conditions understood and the actualities for the eventual transfer of the precious koalas were commenced.

Tama Zoo in Tokyo and Nagoya's Higashiyama Zoo were to receive the koalas from Taronga. Lone Pine Sanctuary in Brisbane was sending Queensland koalas to another Japanese zoo at the same time and while I don't have personal knowledge of their specific preparations, the same export regulations applied nationally.

The most urgent requirement to successfully keep koalas is to have a plentiful supply of fresh eucalypt leaves and Tama and Nagoya Higashiyama Zoos had established eucalypt plantations some years before. So, many months before the planned transfer to Tokyo and Nagoya, shipments of Japanese-grown eucalypt leaves were air-freighted to Sydney on a daily basis for palatability testing by the fussy koalas that were preparing to pack their pouches for the journey to their new home.

A Photographer not a Linguist!

A TRIP TO JAPAN was the prize when Rick Stevens won the Nikon Press Photographer of the Year Award in 1988. Rick, who invariably leaves everything to the last minute, decided the evening before his departure that he should learn at least a few words of useful Japanese. He called in to see me at the Zoo on his way home from work and asked what would be the one word of Japanese he would need most. I told him he would be shown such kindness and would experience such generous hospitality that he would constantly be saying 'thank you', so I suggested he memorise 'arigato' if nothing else. Rick went off muttering 'arigato, arigato' all the way to Tokyo.

The need to demonstrate his Japanese language skills arose the first morning. Rick and his wife, Paula, were endeavouring to negotiate the Tokyo subway system and were obviously looking a little lost. A Japanese gentleman, on the way to catch a train, aborted his own travel plans to kindly show them where and how to purchase tickets, find the right platform and the correct train for the zoo. Now, somewhere in that vast metropolis there is a very confused man probably still wondering what he had done to deserve a very elaborate reference to the gentle art of paper folding? For Rick had bowed deeply to this stranger waving them off from the platform and just as their train was moving off had gratefully and confidently shouted 'ORIGAMI!'.

The importation of the Japanese leaves posed yet another set of obstacles, this time involving Australia's very necessarily stringent quarantine laws. The leaves had to travel in airtight containers and be unpacked in an insect-screened enclosure. They arrived superbly packaged and beautifully labelled and in many different varieties. After partial consumption by the koalas — they only eat the young tips — the remaining branches had to be disposed of as contaminated waste. This was such a shame, as I know the giraffes would have enjoyed sampling the leftover Japanese branches. The usual practice in Taronga is that first the Koalas feed on the juicy tips and the next day the giraffes browse on the rest of the branch. But this was not to be with these imported leaves.

The quality of the Japanese-grown eucalypts was impeccable. Each leaf looked like a large, juicy snow pea compared to the locally harvested, rather 'freeze-dried' gum leaves our koalas were accustomed to. The koalas, now in quarantine awaiting departure, must have thought they were in gum-tree heaven. They devoured the imported leaves with enthusiasm bordering on greed, and any worries about the suitability and acceptance of the eucalypts grown in Japan were dismissed almost instantly. I even saw koala keeper Dave Thomas secretly taste-testing the leaves from Nagoya himself. I half expected him to take some home for dinner they looked so delicious.

The next step was to work with the Japanese zoo directors, keepers and veterinarians on the husbandry and care of koalas. Taronga welcomed zoo staff from Tokyo and Nagoya who stayed for about four or five weeks at a time and immediately found a place in our hearts because of their dedication, their professionalism, their courtesy, their sense of humour, their love of a good party and their willingness to count individual koala droppings morning after morning!

They also provided some early morning excitement for the Zoo neighbours, some of who take a shortcut through Taronga's car park and grounds on their way to catch the ferry to the city before the Zoo is open.

Each day, for weeks, the commuting Zoo neighbours would be greeted by three immaculately khaki-uniformed Japanese gentlemen enthusiastically doing their daily exercises in the crisp morning air outside the Zoo's main entrance. The exercises involved a lot of elaborate posturing and shouting and I think some good burghers of Mosman were a little nervous at first.

Australian wildlife is greatly respected and admired by the Japanese. The zoos receiving Australian native animals are thoroughly investigated by the Australian authorities and the transactions to only the most suitable institutions are government-to-government gifts with all the accompanying prestige and responsibilities. The preparations for the arrival of the koalas in Japan, therefore, were elaborate and the attention to detail was exquisite.

I call the Koala House at Nagoya Higashiyama Zoo the 'Koala Hilton'. The purpose-built, multi-million yen facility has indoor and outdoor display yards, holding areas, food storage cool rooms, a roof that opens on warm days and the Southern Cross displayed on the ceiling so the koalas feel at home at night. Closed-circuit television monitors the koalas' every scratch, munch, poo and fart and the animal behaviour data collected as a result of this careful surveillance is far more extensive than anything available in Australia. The keepers' room in the behind-the-scenes area of the Nagoya Higashiyama Zoo Koala House resembles the cockpit of a jumbo jet with dozens of tiny television monitors displaying the koalas' activity, or inactivity, in the exhibit areas. Similar luxurious conditions for the Koalas exist in Tama Zoo, Tokyo.

The media interest in the early koala shipments was overwhelming at times. Television news crews, journalists, reporters and photographers descended on Taronga for weeks in advance of the departure date. The Japan-bound koalas were in quarantine so access to them was prohibited for very good reasons. All the interviews and filming had to be done in the Koala Walkabout exhibit using Taronga's resident koalas. They looked quite confused at times and seemed to wonder what all the fuss was about. The profile of the koala grew enormously and suddenly the fuzzy, grey, toy-like creatures we all took rather for granted had achieved almost an iconic status.

Hundreds of Japanese media, VIPs, scientists, horticulturists and zoo officials visited us in the months leading up to the koalas' departure and I asked many of them, 'Why do you love koalas?' I only ever received the same answer, 'Because they are so cute'! This was not exactly the seriously considered zoological response I was expecting but it struck a cord with me. They do resemble everyone's childhood toy come to life.

The Japanese Consul General, government Ministers, the Mayor of Mosman, the Mayor of Nagoya, the Governor of Tokyo and even the Premier of New South Wales visited for the farewell party and for photographs with stand-in koalas, as the travelling animals rested in quarantine before departing for their new home.

I often thought that the Taronga veterinarian and zookeeper who were travelling with the animals should have been resting too, but it was not to be. They were in such huge demand for media interviews and preparations for the koalas' wellbeing were so extensive that they barely had time to think about the journey they themselves were about to embark on and what their own personal needs might be. So much so, that we were at the airport

Taronga's Asian Elephants Ranee and Burma ignore the magnificent harbour views to concentrate on breakfast.

daily in Taronga's wonderful Free Flight Bird Show launched in September 1997.
(Photograph courtesy Ben Rushton, the *Sydney Morning Herald*)

The list of recovery and conservation programs for Australian native species carried out at Western Plains Zoo, Dubbo, is long and impressive. The most successful of these is the program for the quiet and unobtrusive Malleefowl.

Bird Keeper Cathy Alexander and Billy the Brolga wander past Taronga's Giraffes during one of Billy's training walks through the Zoo grounds. Billy was a regular and very welcome visitor to the public relations office.

In my opinion, one of Rick's best shots is this one of Giant Pandas Xiao Xiao and Fei Fei at Taronga in July 1988. The long-suffering Xiao Xiao (left) is the somewhat reluctant recipient of an enthusiastic bear hug from playful Fei Fei.

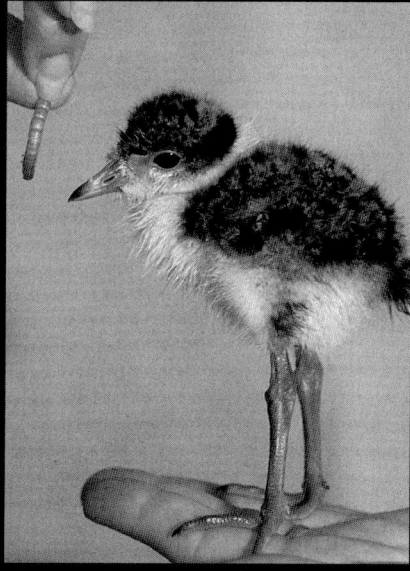

Les Clayton, who retired in 1993 and sadly passed away in 1996, was akin to a Zoo national treasure and tribal elder. He was Taronga's 'bird man' for over 40 years

The birth of Red Panda cubs, usually twins, is always cause for celebration at the Zoo. More than 30 cubs have been born at Taronga since 1976.

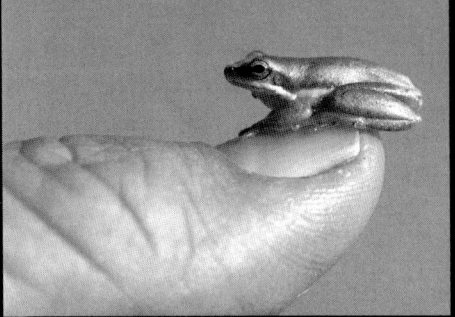

A hitchhiker of the amphibious variety. This Eastern Dwarf Tree Frog, discovered in Sydney amongst a shipment of bananas from Tully, was sent packing back to Queensland.

While being hand-raised by Taronga zookeeper Geoff Ross, Jonathon the Koala travelled around in a backpack. As an adult he eventually moved all the way to Nagoya Higashiyama Zoo, Japan.

Every zoo needs a camel or two, preferably ones that eat hay, not quiche.

My Nagoya Higashiyama Zoo and Nagoya City Council International Relations Division family' at a farewell party in my honour in Nagoya, April 2001. I am sitting fourth from the left in the front row; long-time colleague Eisuke Kashima is on my immediate left

Bruce the Andean Condor almost blocks the sun with his three-metre wing span as he glides across the audience at Taronga's Free Flight Bird Show.

Curator of Birds John DeJose with Diablo (aka Bruce) the Andean Condor chick that was hatched and hand-raised at Taronga Zoo in September 1979.

Leopard Seals in distress, as well as a variety of other marine mammals, find sanctuary at Taronga Zoo every winter. They receive expert veterinary care and eat vast quantities of vitamin-laced fish before being returned to the ocean.

Western Plains Zoo's three African Elephant females Cherie, Yum Yum and Cuddles are long time and high-profile Dubbo residents.

An adult elephant eats about 160 kg of food a day, using its mobile trunk and strong molars to pull down and chew through branches, leaves and grass. Taronga's elephants enjoy stripping banana palms and Moreton Bay fig branches.

Giraffe mother Faye and her calf. The calf was born in September 1987 and named after Rick Stevens who, despite valiant efforts, missed photographing the birth.

Lindy Hippopotamus and her new-born calf, Happy, at Taronga Zoo in 1977. Taronga's Common or Nile Hippopotamuses were all eventually transferred to the bigger ponds and lakes at the open-range Western Plains Zoo, Dubbo.

Przewalski's Horses, bred far away at Western Plains Zoo, Dubbo, and Monarto Zoo, South Australia, take their first tentative steps on the plains of Mongolia in June 1995.

The bold and the beautiful. Designed perfectly for survival on the harsh African plains, not only does the Zebra's striped coat camouflage it, but a new-born foal can stand on its spindly legs within minutes after birth and is grazing with the herd after about a week.

in the departure lounge for only the second such transfer of Koalas when I asked the vet accompanying the animals, David Butcher, where his suitcase was. It was only then that he admitted that he had single-mindedly packed the koalas and the eucalypt leaves and his veterinary kit with every conceivable drug, but had not had time to pack his own bag and was intending to buy some clothes when he got to Japan. Now, he was rather a large veterinarian and I suggested that he might have some difficulty finding clothes to fit him in Japan. As it turned out David had to resort to buying the necessary clothing at a shop for sumo wrestlers and was subsequently known in Japan as 'Sumo-san'.

The jet-setting koalas travelled exceptionally well, and smoothly settled into their new homes. After the regulation quarantine period they went on display and were instant hits with their new and fanatical audiences.

Over the years Taronga and Western Plains Zoos have sent additional koalas to Nagoya and Tokyo to add new breeding stock. The koalas in those zoos and, in fact, other Japanese zoos as well have thrived and bred and there are now more than 100 eucalypt-munching Australian ambassadors sitting in zoo gum trees around that country, being admired and loved by all who look after them and everyone who visits them.

It was disappointing that the Australian media often only reported the deaths of koalas in Japanese zoos, especially in the early days of these animal transfers. This gave a very wrong impression. While some koalas have died, which they do in Australian zoos and in the wild, too, many more have lived long and fruitful lives under the careful watch of their adoptive zookeepers.

One Nagoya-bred koala, a very handsome male called 'Rocks', was presented to Taronga Zoo in December 1993 to celebrate the ten-year anniversary of koalas in that city's zoo. 'Rocks' was a symbol, too, of the successful koala breeding program in Nagoya Higashiyama Zoo.

OUR BONDS OF FRIENDSHIP are especially strong with the staff of Nagoya Higashiyama Zoo. In September 1996 Taronga and Western Plains Zoos signed a formal sister-zoo agreement with that zoo and established a successful and mutually beneficial staff exchange program that continues to develop a greater understanding of and friendship between our two cities, our two countries and our two peoples. This relationship began almost two decades ago when Nagoya Higashiyama's Veterinarian Ono and zookeepers Takahashi and Ando took the first koalas to Nagoya.

Now the sister-zoo relationship encourages an exchange of vital conservation and zoological information and animal husbandry skills. It is an excellent staff development opportunity for people who have the same goals for wildlife conservation.

Much of the success of this relationship is a result of the unstinting input by the Director of Nagoya Higashiyama Zoo, Eisuke Kashima. From almost the very beginning, Kashima has been responsible for the wellbeing of the then koalas in Nagoya, first as their veterinarian and then as Deputy Director and Director of the Zoo and Botanical Gardens. He is also the Studbook Keeper for koalas in Japanese zoos but much more than this, Kashima recognises the importance of the koala as an ambassador for international friendship and as a symbol of the need for wildlife conservation.

Kashima, his staff and Zoo volunteers and the past and present members of the international division of Nagoya City Council have given me so much support over the years. They have become my

> ## TREETOP VEGETARIANS
>
> KOALAS EAT MANY EUCALYPT and occasionally some non-eucalypt species. Their main food trees are River Red gum and Forest Red gum in the north and Grey, Manna and Swamp gums in the southeast. The leaves have a high moisture content so koalas only drink water very occasionally. Koalas have also been known to descend from the trees and eat soil and gravel.
>
> Species of eucalypt favoured by Taronga's koalas: River Red gum, Manna gum, Forest Red gum, Scribbly gum, Tallowwood, Narrow-leafed Ironbark.
>
> Quantity of browse collected for Taronga's koalas annually: 24,000 eucalypt branches (to feed 22 koalas)
>
> Eucalypt collection sites for Taronga's koalas: Zoo Botanic Estate staff still collect eucalypt branches from the trees planted in northern beaches schools over 20 years ago. The primary collection location is, however, a 1.8 ha plantation of 4,000 eucalypt trees at the University of Western Sydney, Hawkesbury campus.

second zoo family and I count myself very fortunate to have this greatly valued Nagoya 'stitch' in my colourful zoo tapestry.

I am proud to be a member of the Sydney-Nagoya Sister City Committee, which is part of the City of Sydney's International Affairs Program, and I hope to be actively involved for a long time to come.

In November 2000, Zoo staff joined Sydney and Nagoya citizens at a celebratory lunch and tree planting ceremony at Taronga to mark the 20-year anniversary of the sister-city relationship. It was a happy and memorable occasion, and was one of my last official Zoo events I was honoured to share with so many international guests who have such a passion for Australian wildlife.

Zoo business has taken me to Japan — Nagoya in particular — numerous times since 1984, and I am constantly delighted and encouraged by the high level of interest there in our beautiful city of Sydney. The citizens of Nagoya have an extensive knowledge of the Sydney sister-city relationship, an enthusiastic interest in either visiting or re-visiting our country and a passion for our unique Australian wildlife. I am firmly convinced that the koalas in Nagoya Higashiyama Zoo are responsible for much of this positive profile.

I was delighted to learn that a female koala, which went to Nagoya Higashiyama Zoo from Taronga in September 2001, was named 'Clements' and that the next male koala to go there will be named 'Darill'. I know that these animals, like their namesake, will be welcomed with great kindness and will receive wonderful care in Nagoya. I am confident that these koalas, as with their predecessors, will do much to nurture the existing bonds of friendship between Sydney and Nagoya and be outstanding ambassadors there for wildlife conservation for years to come.

CHAPTER SEVEN

85 Years Caring for Wildlife

As well as providing exceptional care for their own animals — on average a total of almost 3,500 mammals, birds, reptiles and fish — Taronga and Western Plains Zoos every year provide sanctuary for more than 1,200 sick, injured or orphaned native animals brought to the Zoos by members of the public for treatment. The interest that the general public have in wildlife issues and their care for native animals in distress is very encouraging and I have seen a huge attitudinal change in the community over the years. Many people go to great lengths to assist animals in trouble. They pick up animal accident victims and often bring them long distances for the specialist veterinary care the Zoos can provide.

The Zoos' present veterinarians, Larry Vogelnest, Julie Barnes, Francis Hulst and Ben Bryant, the wonderfully dedicated veterinary nurses and Taronga's tireless Wildlife Rehabilitation

Supervisor Elizabeth Hall, like all their predecessors, go to amazing lengths to patch up these critters and return them to a life in the wild whenever it is possible.

This resource-consuming Zoo work provides a much-needed public service and encourages the community to get involved in caring for wildlife, which can only be to our native species' advantage. The rescue, rehabilitation and release programs for these sick or injured native animals often involve whole families, schools and communities and provide powerful environmental lessons for all concerned.

For a quarter of a century I was able to interest the media in a myriad of these fascinating animal rescue and release stories and could almost always guarantee memorable images and happy endings, so it is very difficult to single out just a few special cases. It is more like what can I leave out rather than what should I include but I will do my best to provide a broad cross-section of species, surprises and special situations.

FORGET CROCODILE DUNDEE as the most recognisable international image of the outdoors, bronzed Aussie male. In the late 1970s, television footage taken at Taronga by a Japanese broadcaster included some zookeeper antics that would have made Mick Dundee look like a junior boy scout.

Japanese media visited Taronga often and were fascinated with images of and information on our Australian native animals. It was always an excellent opportunity to promote this country's unique natural heritage and to get some conservation messages out to a very large and significant audience.

The Zoo's Wildlife Clinic, while providing sanctuary for many

> ## Ringtail Possum Facts
>
> THE RINGTAIL POSSUM (*Pseudocheirus peregrinus*) faces many perils in coastal regions of Australia where it lives, often close to human habitation, from the tip of Cape York down to Tasmania with a further sub-species in the far south of Western Australia. This small, very pretty marsupial measures up to 350 mm in length and has a prehensile tail with a white tip or end, which has a friction pad on it, which is often used as a 'fifth limb'. The Ringtail Possum is usually rufous red in colour and builds a complex spherical nest or drey. It has become quite adapted for life in suburbia and eats a variety of suburban garden plants including rose petals, much to some gardeners' dismay. They don't seem as wily or streetwise as their larger Brushtail Possum cousins and they often fall victim to cat and dog attacks and to speeding cars.

animals recovering from trauma, invariably could provide close-up pictures of mammals, birds and reptiles, all of which came with interesting stories or histories. The animals were generally easy to handle or to get close to, so I usually made the veterinary hospital the first stop on our media tour.

One morning in December 1978, we arrived at the hospital to meet up with enthusiastic young veterinary keeper Geoff Kidd who had begun working at Taronga as a teenager about a year or so after I started. As is the usual way with the Japanese media, they arrived about 30 minutes early, which meant we caught Geoff still busy with his early morning cleaning, hosing and feeding duties.

There he was dressed only in his khaki work shorts and long black gumboots, frantically sweeping and hosing in an effort to have everything shipshape by the time we arrived. The Japanese, especially the female television presenter, liked what they saw and the cameras came out immediately. Geoff is forever helpful and before I knew it he had put away his cleaning equipment and had brought out a young Ringtail Possum for inclusion in the filming. The possum had been the victim of a cat attack but, after treatment in the veterinary hospital, had made an excellent recovery and was almost ready for release.

Topless Geoff must have noticed me glaring at him for the obvious breach of Zoo uniform policy as he was desperately trying to hold still the squirming possum for the cameras with one hand and grab his shirt with the other when all of a sudden, the frisky television star made a leap for freedom and ran up the nearest gum tree.

Before I knew it, bare-chested Geoff was up the tree after the escapee. The possum obviously felt extremely well and began leaping from one tree to another with Geoff in hot pursuit, looking exactly like Tarzan of Mosman. Blond hair, suntanned bare chest, short shorts — all that was missing was the primeval yell as he swung from tree to tree, with the beautiful views of Sydney Harbour as the backdrop. This is the sort of thing international television producers must dream about and I am sure the broadcast did wonders for Australia's tourism promotion in Japan. I am also sure that Geoffrey, 20 years on, still gets fan mail from Japan!

Geoff would never forgive me if I failed to record that 'Tarzan of Mosman' eventually caught that very energetic Ringtail Possum which, after completing its scheduled convalescence, was released into the safety of bushland near Manly Dam — this must have seemed all very dull in comparison with its previous adventures.

IN THE WINTER OF 1980, a Wandering Albatross was brought to Taronga's Veterinary Clinic having been found with fishing line caught around its neck. Sadly, this is a very common occurrence for sea birds of many species and the huge albatross seems to fall victim more than most.

The Wandering Albatross (*Diomedea exulans*) is the largest flying bird alive today and an adult has a wingspan of nearly three and a half metres. It wanders the globe in southern latitudes and can live to more than 40 years of age. It is majestic in flight and has beautiful white and grey plumage, the edges of which are finely marked with brown to give a beautiful scalloped effect. It has a large, pink bill, which does a lot of 'clappering' during courtship.

Sadly, Wandering Albatross numbers are declining rapidly in the wild. Many get caught up in fishing line and nets and organisations like the Southern Oceans Seabird Study Association are working hard to secure a future for these giants of the ocean waves.

It is vital that large sea birds spend as little time as possible in captivity, where they can be prone to contracting foot and lung problems, if they are to be rehabilitated successfully and released to stand any chance of subsequent survival in the wild. It is always a delicate balancing act and three weeks seems about the optimum time for recuperation.

Taronga's veterinary team worked fast and hard to mend this particular Wandering Albatross and fatten it up to peak condition so it had the very best chance of survival on its release.

I love Libby Hall for many reasons, but probably mostly because she seems to always dream up appropriate 'stage' names for her charges! I might need to retract that statement as I have just remembered she recently called a rather dried-up looking Shingleback Skink, Dazza, after me! She, however, gave this Wandering Albatross the romantic name of Anastasia. Anastasia responded well to veterinary treatment, swallowed her antibiotics and vitamins like a good bird should and generally thrived on a handfed diet of fresh seafood such as yellowtail and squid.

Soon Anastasia was ready to return to a life on the ocean waves so Curator of Birds John DeJose, with the help of Libby Hall, fitted her with an identifying leg band for possible post-release monitoring, packed her into a crate, and drove her to her release point — the windswept cliffs of Sydney Harbour's North Head. This picturesque headland above Manly marks the northernmost entry point to our beautiful harbour and was sure to be a perfect launching pad for this albatross's release. In the meantime, I packed up a large media contingent and we formed a convoy to North Head to farewell this magnificent bird.

We drove to the national park, left our cars and carried the crate containing the albatross as close to the edge of the cliff as possible. John DeJose explained to the gathered media that he planned to open the crate door and let Anastasia come out to take off in her own good time. These large, heavy birds should benefit from the updrafts from the cliffs for take-off, and on this particular August day the winds from the Southern Ocean were cool and strong — perfect conditions for a Wandering Albatross.

The door was opened but Anastasia failed to put even her beak out. We sat back and waited and waited.

After about two hours, Anastasia decided to waddle out of the crate to the cliff edge, but instead of facing out to sea to contemplate departure she turned around to face the assembled media. I couldn't help thinking that she was also facing in the direction of Taronga Zoo and was perhaps fondly remembering all that fresh fish which was handed to her every day.

Media deadlines came and went as we waited even longer for Anastasia to lift off. But it was not to be. John and Libby decided that the wind was probably coming from the wrong direction and that the best plan would be to put the bird back into the crate, return to the Zoo and try again the next day in the hope of a change in wind direction. The media were all very patient and understanding and promised to regroup in the morning at the same time and location.

The following day dawned bright and clear with more amenable offshore winds so there we were again, all back up at North Head for an action replay. This time, Anastasia walked immediately out of the crate but again sat on the edge looking down to the rocks far below with a very strange look on her face. It was almost as if she was afraid of heights. She kept looking down to the crashing waves below and turning away again very quickly. Nothing seemed to be enticing her to make a getaway.

After about another two hours of albatross watching, John DeJose called me aside and said he thought he would have to throw Anastasia off the cliff as he was convinced that she would probably never make the move herself. I went pale at the thought, but he assured me that once she caught the updraft she would spread her wings and be off.

Feeling none too confident, I explained the plan to the media who prepared themselves for action. I am sure some of them felt they would be covering a disaster rather than a success story and I think I felt a little that way myself, but I tried to remain optimistic.

John gathered the huge, strong bird into his arms, walked right to the edge of the cliff, raised her above his head and threw her as far up as he could. To our dismay she went straight down like a stone and disappeared below the cliff edge. My heart was in my mouth as we waited for what seemed like an eternity before we all dared to look. Then up she soared on the strong thermals and with her massive wings spread, she effortlessly headed south.

There was a huge cheer, many tears, and wonderful images for the news bulletins that evening and front-page picture stories the next day.

I have to confess now, though, that Anastasia had not read the wording of my news release, and did not immediately 'fly off to a life on the ocean waves' but was found on Bondi Beach a couple of days later. The band around her leg identified her as our recent visitor and she was brought back to the Zoo. She was checked over, found to be in good health but was given the benefit of the doubt and enjoyed another few days rest, recuperation and more fresh yellowtail.

Libby gave her some more flying lessons just to make sure her wings were in good working order and then enlisted the assistance of the Royal Australian Navy. The Navy kindly took the albatross and Libby five kilometres out to sea on one of their small boats so Libby could launch Anastasia directly onto the ocean and watch her skim the waves until she disappeared out of sight.

I'll long remember Anastasia, the albatross that was afraid of heights. Her legacy at the Zoo has been that all albatross releases are now done from a boat and not a cliff top.

LITTLE PENGUINS, COLONIES of which exist in and around the Sydney waterways, often find themselves the victims of dog attacks, oil spills, fishing line or hook injuries or some other urban peril and end up spending time in Taronga's care. Despite

> ## PENGUIN FACTS
>
> LITTLE PENGUINS (*Eudyptula minor*) have also been known as Blue or Fairy Penguins. Their bodies, about 330 mm long, are like little torpedoes. They have short, thick necks and their small, thick feathers lie flat against their body overlaying a layer of down which entirely waterproofs them and provides exceptional insulation.
>
> This species is the smallest of all the penguins, is the only one to breed in Australia and the only one to wait till dark before coming ashore to roost in burrows, crevices and under tussocks. Just before dawn the birds converge on pathways to the beach and waddle quickly into the water. They 'fly' underwater, flapping their modified wings or

what may often seem like insurmountable problems, these quaint-looking but quite robust birds respond well to treatment at the Zoo and more often than not are able to be released back to sea.

One of the most satisfying aspects of my public relations work at the Zoo was being able to accompany Libby Hall and her team, along with the media, on numerous penguin release excursions. Like the albatross, these birds are also fitted with a leg band for future identification, and quite often four or five Little Penguins at a time are loaded into pet-packs and driven to one of the quieter northern beaches for release. They are never reluctant to depart and always scurry enthusiastically into the waves without so much as a backwards glance.

IN LATE WINTER EVERY year seals begin their journey south, following the cold currents back to Antarctic waters for summer. Along the way a number of them always seem to come to grief in some way or other. Often they have deep wounds from Cookie-cutter Shark attacks, or deeply imbedded stingray spines from attempting to catch the wrong kind of food. Sometimes they have been caught in fishing nets, have swallowed plastic bags thinking they were jellyfish or, in the case of the young seals, have been separated from the adults and are exhausted and hungry.

Taronga Zoo receives numerous calls to assist sick, injured or just tired seals found stranded on the beaches of Sydney. These marine mammals, usually the very young or slow and elderly, seek refuge in the sheltered bays and beaches up and down Sydney's coast.

Taronga's veterinarians and keepers work closely with the New South Wales National Parks and Wildlife Service rangers

NOT SO LITTLE PENGUIN

THE LITTLE PENGUIN BEACH at Taronga was always a great location for filming and for television interviews. The residents of that exhibit are quite friendly and cooperative, particularly at meal time but they almost bit off more than they could chew when reporter Jonathan Coleman, dressed in a giant penguin suit, arrived to record a story for the first episode of the long-running children's television show, 'Simon Townsend's Wonder World', way back in 1979. Jono, a crazy man from day one, ended his story about Fairy Penguins by doing a giant belly flop into their freezing pool. The large suit meant Jonathan sank straight to the bottom of the pool and the penguin residents looked amazed as the intrepid but bedraggled reporter eventually hauled himself out. In an effort to revive the shivering Jono, I quickly made him a cup of hot coffee at the nearby Seal Theatre. However, instead of reviving Jonathan I almost poisoned him as I had mistakenly used water from the saltwater tap!

and monitor these seals' condition and patterns of behaviour. Often the same animal will be seen for weeks on different beaches around Sydney and the community is very helpful in reporting such sightings. The seal may just be in need of a long rest and often spends time basking in the sun before disappearing to continue on its way further south.

The Zoo is called in if a seal has very obvious signs of illness or injury and sometimes it is vital that these animals be brought back to Taronga's marine mammal pools where they can have intensive veterinary treatment. While the plan would be to ultimately rehabilitate and release such an animal whenever possible, picking up a seal off the beach is always a very sensitive issue and some members of the community insist the animal would be better left where it is found. I have witnessed some very difficult scenes on the beach as well-meaning but often misguided members of the public argue with the National Parks officers and Zoo vets while the seal lies suffering at their feet.

Young Leopard Seals a long way from the Antarctic pack ice often fall victim to shark attacks and many of these huge, spotted marine mammals with the most beautiful brown eyes and the sharpest teeth imaginable end up in need of rescue. The Leopard Seal (*Hydrurga leptonyx*), also known as the Sea Leopard, feeds extensively on warm-blooded animals such as penguins but it also eats krill in large amounts and may also feed on carrion and fish. It is also known to attack other species of young seal. It is a solitary animal, inhabiting the outer fringes of the Antarctic pack ice. This large marine mammal grows to a length of up to 360 centimetres and can weigh up to 450 kilograms. The females are bigger than the males.

Once Taronga receives a call for assistance from the National Parks Service to assess, treat and possibly take in the injured seal, a Zoo truck equipped with cargo nets and a large stretcher, several keepers and a veterinarian set off for the stranding site. If luck is on their side, they are called to a location quite close by but sometimes it involves a long drive north or south of Sydney.

I vividly recall one of my earliest involvements in a seal rescue. We were called out to Garie Beach in the Royal National Park about an hour and a half drive south of Sydney. When we arrived a large crowd of onlookers was gathered on the cold and blustery beach and several members of the media were there as well.

The Leopard Seal in question had hauled itself up on some rather inaccessible rocks at one end of the beach. It was very obviously injured but still feisty. The National Parks Service rangers had been monitoring the adult animal for some time and had decided that it would not recover unless it had some veterinary assistance and some 'R & R' at Taronga.

As I said, these Leopard Seals are large, strong and aggressive and this one, despite being far from peak condition, could still inflict a very serious injury if anyone were bitten. The first thing we needed to do in the rescue process was to secure the seal in a large, strong cargo net before it was placed onto the stretcher for closer veterinary inspection. Elementary treatment is often given prior to transfer to the Zoo.

Big, strong John West, the Zoo's senior Operations Manager of considerable experience, expertise and a very dry sense of humour is a great asset in such situations. He is a clear thinker, determined, focused and always gets the job done, come what may. This particular day he had to battle a 350-kilogram Leopard Seal and a somewhat hostile crowd to succeed in his task.

The last thing this seal wanted was a cargo net thrown over it and it made a very determined bid to escape. Despite its injuries, the net and the fact that big John and three other keepers were holding onto it, the seal lurched forward at considerable speed across the jagged rocks towards the sea. The exhausted keepers lost their grip as the seal got closer to the water — all except John West who was hanging onto the back flippers of the animal and being dragged across the rocks to cries from the onlookers of: 'Be careful of the seal! Don't hurt the seal!' Finally, John's strength and determination won out and the ungrateful seal was loaded safely onto the truck for its trip to the Zoo veterinary hospital. The look on John's face as he surveyed his shredded wet suit and chest lacerations said: 'Forget the bloody seal, what about me!'

I should add that both the Leopard Seal and John recovered to fight another day. The seal was subsequently rehabilitated and released, and John West is still doing good things for the animals and staff at Taronga and, when the need arises, still rescuing these less than grateful seals.

OFTEN THE ZOO HAS to go to great and complicated lengths to secure a happy ending for displaced creatures who get themselves into difficulties. Many different individuals and organisations come to our rescue in times like this and help Taronga return these native animals to their rightful place in the wild.

The Royal Australian Navy, local and interstate airlines, truck drivers and ordinary caring individuals come to the Zoo's aid with offers of transport of all kinds.

The Royal Australian Navy once transported two Green Turtles (*Chelonia mydas*) back to the warm waters of North Queensland.

The particular turtles I am recalling were still small, about the size of a saucer, and had turned up on a Sydney beach in the middle

> ## GREEN TURTLE FACTS
>
> OF THE SEVEN SPECIES of marine turtle in the world, six are found in Australian waters. The Green Turtle is Australia's most common, and is found in tropical waters, especially around the Great Barrier Reef. It is reasonably safe and numerous in Australia but in other countries it has virtually been exterminated because of indiscriminate exploitation of its prized flesh, oil and skin. In Australia, it is a protected species but natural predators do take a heavy toll. When the baby turtles hatch they often fall prey to birds when making their fumbling dash from the nest site to the sea. If they reach maturity, an adult Green Turtle can weigh up to 180 kg and grow to about a metre long. This turtle feeds on algae, marine plants and occasionally on jellyfish. The young are more carnivorous than the adults.

of winter. They were cold, dehydrated and hundreds of kilometres from home. We could only guess how the youngsters got there but perhaps, because they are very appealing and cute, they may have been picked up in Queensland by tourists and brought back as possible pets. When they wouldn't eat they may have been returned to the sea in the vain hope they would survive on their own.

Fortunately for the little turtles they had somehow been washed up in the surf, were picked up by a passer-by and ended up in a bathtub of warm, salty water at Taronga. Happily for these two turtles and another larger one, which had been rescued separately, the Royal Australian Navy had a PT boat heading from Sydney to Cairns on exercises and assured me they would be happy to transport the unusual cargo. After veterinary treatment and plenty of vitamin-laced fish, the three turtles were loaded into fibreglass transport ponds and transferred to the naval vessel. After a somewhat official naval farewell we waved goodbye to the turtles, confident that once they were off Cairns they would be lowered into the warm northern Pacific Ocean where they belonged.

I received a telex from the boat's captain about a week after their departure advising me that the turtles had been returned home safely. The telex also noted that the threesome had faced north for the entire sea journey. Obviously the turtles' internal compasses were letting them know exactly where they should be headed or maybe they were considering a career in the Navy and were just showing off their innate navigational skills.

SOME FROGS HAVE a particular penchant for getting themselves into all kinds of bother and seem to regularly plan trips south for a holiday. Frogs are nocturnal and in the daytime rest in damp, shady places safely hidden among thick plants or under rocks or logs. The frogs of Tully, the wettest place in far North Queensland, live in the banana plantations, finding shelter under the large green banana palm leaves. They have good night vision but are not so good in the day, and obviously don't see the banana crop cutters coming.

Many a frog such as the bulky Green Tree Frog (*Litoria caerulea*) which can grow to about 100 millimetres in length, or the more diminutive Dainty Tree Frog (*Litoria gracilenta*) measuring about 45 millimetres, has been inadvertently packed into a cardboard box along with Queensland's finest bananas, put into a refrigerated truck and driven on a long and shivery journey all the way to the Sydney Fruit Markets some 1,800 kilometres from home. Sometimes the frogs make their presence felt when the cartons come out of cold storage at the markets. Other times the amphibian hitchhikers travel all the way to a Woolworths or Coles supermarket in an outlying Sydney suburb before they are discovered amongst the fruit. Imagine the chaos these croaking little green creatures cause on the supermarket shelves.

Many of the people who discover the frogs very kindly bring them to the Zoo where they stay in quarantine until we can hopefully arrange a free return flight or truck journey for them back to Tully. Libby Hall can usually pinpoint the exact plantation the hitchhikers come from because of the well-labelled banana boxes. She works closely with the Quarantine and National Parks Authorities and I am sure these frogs have a lot of tales to croak about once they return home.

ONCE OR TWICE LIBBY HALL presented me with a rescue, rehabilitation and release predicament of particularly thorny proportions, which required me to convince an airline to carry a very strange passenger by the name of *Moloch horridus* on a one-way flight from Sydney to Alice Springs.

Moloch horridus, also known as the Thorny Devil, is an extraordinary-looking lizard from the deserts of the central and western parts of Australia. It grows to about 15 centimetres in length and has tiny legs, lots of lumps and bumps, blotchy red and cream colours, sharp spines, a short tail and, to some, appears quite grotesque. Thorny Devils can, it seems, be particularly repulsive if you find them in your sleeping bag once you have returned to Sydney from a camping holiday in the outback. One particular Thorny Devil came to light in the bedroom of a happy camper who thought she had returned from holidays with an alien as a souvenir.

The woman, still in a state of shock, brought the little stowaway to Taronga's wildlife clinic. A daily diet of thousands of black ants was required to keep the little devil alive until it could be given a veterinary checkup and the necessary transport arranged for its return to its place of origin.

I pity Libby Hall every time such unusual creatures arrive. Not only do they require extremely complicated care and attention while they are at the Zoo, but the organisation required to return them to their place of origin and release them safely rivals that of a complex military exercise. To Libby and the team in Taronga's veterinary clinic, nothing is too much trouble for an animal in distress.

Invariably, I was able to convince one of our domestic airlines to take *M. horridus* on board. The thorny passenger usually travels in a small Esky box in the cockpit of the plane so as not to frighten the paying customers. At Alice Springs a National Parks Service officer picks up the erstwhile traveller and drives it as far away from civilisation as possible. It is then released onto a pile of unsuspecting ants and hopefully remembers never to follow a trail of such ants into a sleeping bag again.

ANOTHER ANIMAL THAT REGULARLY finds itself in difficult situations and often finds its way to the wildlife clinics at Taronga and Western Plains Zoos is the Common Wombat (*Vombatus ursinus*). Wombats are extremely deceptive creatures and I certainly would never like to find one in my sleeping bag. I also learned early in my Zoo public relations career never to trust a wombat.

Most people think of this distant relative of the Koala as a muddle-headed, cartoon character that blunders its way through life looking a lot like a cuddly, soft toy. In fact, I have seen a wombat chase a keeper so she practically leapt a tall building in a single bound to escape. I have seen a wombat inflict a shin wound so deep it required ten stitches, almost ruin a wedding and attempt to demolish Taronga's Education Centre building. I guess when you can weigh up to 30 kilograms and grow to about 114 centimetres in length, have a broad, blunt head, exceptionally large and strong teeth, powerful forepaws, long claws and a backside that is reinforced like an armoured truck, you could do a fair amount of serious damage to most things if you set your mind to it.

One such wombat I remember very vividly was called Muswells (pronounced muscles), an orphaned Common Wombat found on the roadside near the New South Wales country town of Muswellbrook. His mother had been killed by a car and a passing motorist had stopped to check the pouch of the dead animal and found a tiny pink 'piglet' curled up inside, unhurt. She hand-raised this animal for some time but then decided that Muswells's future was in the Zoo, and contacted Taronga's Education Centre to see if we could take him in. I am sure he was a handful even at that young age and perhaps his surrogate mother was very glad to be rid of him. He had a temperament and build to match his name, but he seemed to relate better to people than animals so maybe was not a candidate for eventual release.

Muswells took up residence as a classroom animal in the Zoo school. Injured or orphaned native animals which are brought to the Zoo for care and which, for a variety of reasons, cannot ever be released, often enjoy long, happy and useful lives helping the education officers teach visiting school children about wildlife adaptations and conservation issues. Over 100,000 school children visit Taronga and Western Plains Zoos annually and about 25,000 of them have a lesson in the Zoo Education Centres. Today's children are tomorrow's decision makers, conservationists and protectors of wildlife, so the Zoos' environmental education role is vital and these animals are often the star players.

Muswells developed into a strapping young male and when he grew too large for his classroom enclosure began sleeping in the staff room shower or toilet until he was needed in class. He frightened many a visitor to the Education Centre, including myself, by insisting on sleeping at the base of the toilet bowl. It is not easy negotiating a snoring wombat with your trousers down around your ankles. I, for one, was very glad when Muswells eventually moved on to another zoo.

Government Ministers who are responsible for our State's two wonderful zoological institutions usually make the most of the very different and useful media opportunities regularly presented by Taronga and Western Plains Zoos. The announcement of a new Zoo director invariably brought the Minister across the harbour for photographs with the new Zoo boss and, of course, with some of his animals.

One windy August afternoon in 1978 all the elements were there for an excellent publicity opportunity for the Zoos and the Minister. Muswells the wombat was part of this scenario. I had also organised for a young, hand-raised Red Kangaroo to be

Upstaged by a Wombat

I ALMOST MADE a friendship-limiting move when I suggested to the Education Centre Manager, Deborah Vitlin, that she should have Gloria the wombat make an appearance at her upcoming wedding, which she was planning to hold at the Zoo. I thought Gloria, an unusually well-mannered wombat who was very much part of the Education Centre 'family' in 1980, would make a unique 'bridesmaid' for Deborah and provide a very special Zoo touch to the happy occasion. The celebrations were proceeding very smoothly until the wombat arrived and grabbed the limelight. From the moment Gloria appeared, the bride and her groom, James, were forgotten as the wedding guests queued up to have their photos taken with the furry bridesmaid. Deborah is still my friend so I don't believe she was too damaged by being upstaged by a wombat on her wedding day.

present so the Minister could hold the wombat and new director Jack Throp, who had recently arrived from Honolulu Zoo, could get a warm and welcoming cuddle from the kangaroo.

After the official announcement by the Minister Bill Crabtree, the media asked for the much sought after photographs with animals and in came Muswells and the little Red Kangaroo on cue. Unfortunately and unbeknown to me, the wombat and the kangaroo had never met each other before. While the kangaroo joey was impeccably behaved, the wretched wombat took an instant dislike to this little upstart with the long tail, big feet and bounce in his step.

With his lethal teeth at the ready, Muswells did a torpedo-like lunge towards the unsuspecting kangaroo, which, very sensibly and speedily, hopped out of harm's way. The menacing wombat teeth missed the kangaroo and sank deeply into the ministerial shin.

There was blood everywhere, and afternoon tea was forgotten as the Minister was rushed to the local medical centre for stitches and tetanus shots, and the media rushed off with their front-page pictures.

I now knew why the Education Centre secretary, Lorna Brennan, wore long leather boots even at the height of summer, and typed her letters with her feet up on a chair — she would do anything to avoid the wombat menace.

It was a career-limiting move for Muswells. The last thing I heard he was packing his bags for a zoo somewhere in Germany. I was never sure what kind of international ambassador he made. I do know that for years after the Minister took great delight displaying his war wounds and recounting the tale of how he took on Taronga's killer wombat.

I SOMETIMES WONDERED whether word got around, particularly in the bird world, that Taronga Zoo was a soft touch for a rest, plenty of care and attention, fine food and a warm nest. Often, it seemed to me, injured birds went thousands of kilometres out of their way to get 'rescued' by a Taronga veterinary nurse or keeper.

One such bird was a Tahiti Petrel (*Pterodroma rostrata*) that was found covered in oil in Sydney Harbour in December 1998. According to *The Reader's Digest Complete Book of Australian Birds,* this is one of the least-known of the petrel species. It is described as a gadfly petrel, which is rarely seen in eastern Australian waters. It nests on no more than a handful of islands in the South Pacific and the Indian Ocean, over summer and autumn. It is a tireless flier and only visits land when nesting, and then only at night. Its closest nesting site to Sydney would be the islands of New Caledonia.

This particular Tahiti Petrel had probably been covered in oil at sea and, weak and unable to fly, had very fortunately landed on a passing ship, which was on its way to Sydney. Once in Sydney, the uncommon visitor to these shores was discovered and brought to the Zoo's clinic where it stayed for several weeks while it was cleaned of the terrible oil and its feathers became waterproof once more.

Towards the end of its recuperation period I used to visit the clinic regularly to discuss with Libby the possible transport options for the bird. Despite the fact that it had made a wonderful recovery from its ordeal, there was no way we could release this bird in Sydney and expect it to fly all the way back to Noumea.

Libby was usually running around the recovery aviary giving the bird physiotherapy and flying lessons that involved her hanging onto the bird's feet and moving it up and down as it flapped its wings madly. I often wondered who was getting the most exercise — the rehabilitation supervisor or her patient?

My husband has worked with the shipping company, P&O Cruises, for over 30 years and as a result I have enjoyed great support from them for many a weird and wonderful Zoo cause.

Recommendations from the Zoos' NRMA Wildlife Clinics

HOW CAN MOTORISTS HELP?
- Drive carefully at dawn and dusk when many animals are active.
- Check the pouches of dead animals for joeys that sometimes remain uninjured.
- Don't throw food from cars as this attracts animals onto the road.
- Remove dead animals from the road as they attract meat-eating animals that may then be hit by a car.

First Aid for Injured Animals

- Place an unconscious animal on its side.
- Keep the animal warm and quiet. Small orphans (pouch young) can be wrapped up and placed inside clothing to provide warmth.
- Stop bleeding with a pressure bandage.
- Don't try to force the animal to eat or drink.
- Keep the animal away from young children, domestic pets and loud noises.
- Don't handle more than necessary.

> ## Transport
>
> ♠ Transport animal to Taronga or Western Plains Zoos' NRMA Wildlife Clinics, a veterinarian or wildlife carer as quickly as possible.
> ♠ Place the animal on its side on the floor of the vehicle (if immobile) or in a cardboard box with lots of ventilation holes and towels; wrap it in a towel or place in a pillow case.
> ♠ Minimise noise (turn off the car radio).

In fact, I am sure their marketing manager cringed every time he heard my voice on the phone and must have thought: 'What on earth does she want now?'

Cruises have been generously donated as prizes for Zoo fundraising efforts and a P&O ship has transported environmental education books to Suva for us for distribution to Fijian school children to involve them more in saving the endangered Fijian Crested Iguana. Once an outboard motor and a small boat for the ranger responsible for monitoring the rare iguanas on the remote Fijian islands were shipped to Suva courtesy of a P&O liner. On this particular occasion in December 1999, P&O very willingly offered to transport Libby and the Tahiti Petrel on board their ship *Fair Princess* to the warm waters off Noumea. Both Libby and the bird were delighted.

The ship's skipper, Captain Philip Pickford, turned out to be a bird enthusiast who insisted the pet pack containing the rather smelly, fish-eating petrel travel on the bridge with him and the ship's officers. He ensured that the passengers were all aware of the unusual cargo on board and every day gave public address and newsletter announcements on the petrel's progress and condition.

The day before arriving in Noumea, Libby sighted other Tahiti Petrels skimming the ocean waves and decided this was the place to release her little charge. Hundreds of passengers gathered on the back decks of *Fair Princess* as Libby opened the door of the pet pack and the Tahiti Petrel flew free once more. The passengers, some of them in tears, watched intently until the bird was merely a tiny speck on the horizon. I am sure that witnessing this unusual and enriching event was one holiday experience these cruise passengers were not expecting.

CHAPTER 8

Pandamonium

THE AUSTRALIAN CONCISE Oxford Dictionary describes pandemonium as *'abode of all demons; place of lawless violence or uproar; utter confusion'*. While Taronga was not quite, and I trust never will be, an abode of all demons or of lawlessness and violence, the months leading up to the visit of the Giant Pandas to our Zoo in winter 1988 were often utterly frantic, confusing and uproarious.

It was at that time that we decided that the word 'Pandamonium' was the very best way to describe and publicise the long-awaited and much-heralded arrival of the two black-and-white fluffy creatures from China.

These Giant Pandas were the official gift to the people of Australia from the People's Republic of China for this country's bicentennial celebrations in 1988. It was the first and quite possibly the only time these rare animals, such high profile ambassadors for the world's endangered species, would be seen in our country. The pandas, on loan only, would first visit

Melbourne Zoo for three months and then arrive at Taronga Zoo in July 1988 for a further three months.

Zoo preparations for the visit of the Giant Pandas began about 18 months before their anticipated arrival. It took months of work by many to plan and build a suitable exhibit, train keepers in panda husbandry, ensure adequate bamboo supplies, organise ticketing, parking and public transport arrangements, recruit additional security and customer service staff, plan the vital education, publicity and public relations campaigns and to make sure we were in a position to take full advantage of the associated retail and catering opportunities the panda visit would present.

The complex political negotiations to secure this first-ever visit to Australia of one of the world's most rare animals took almost seven years. It was a coup for our country and our Zoo chairman at that time, Mr Ted Harris. These sometimes painful planning and negotiating processes — which required patience of giant proportions over a long period of time — disappeared from our immediate memories once these most charismatic of animals arrived in our midst.

In planning for this exciting event, our objectives were to exhibit the Giant Pandas in a way that would ensure optimal conditions for these special animals and provide great enjoyment and satisfaction for the visitors to the Zoo. It was imperative that we take full advantage of this unique opportunity to educate the community on a broad range of conservation issues.

A new exhibit was constructed which featured a moat separating the animals and visitors, climbing logs and a sandpit, shade trees, a stream and landscaping. There were even fine mist sprays that would keep the pandas comfortable during warmer weather. In addition, visitors would view the pandas from an elevated timber walkway and behind-the-scenes there were heated night-dens complete with timber futons for the animals and state-of-the-art food preparation facilities for the keepers. For security purposes, there were viewing windows from the kitchen area into the pandas' night-dens as 24-hour surveillance by security staff was to be in place while the precious animals were in our care.

Taronga's resident elephants Heman, Ranee and Burma watched in amazement as a huge red Chinese welcome gate was built near their temple home. They probably thought it was a special decoration for them, but it was actually to mark the beginning of the queue route to the Giant Panda viewing platform. Along this pathway was a series of fabulous interpretive graphics designed to give visitors more information about the fascinating Giant Panda, its habits and habitat, behaviour and physiology, its endangered status and what people could do to help save this species. These graphic panels as well as numerous touch tables, decorative banners and all kinds of panda-inspired children's activities had been developed to celebrate the much-heralded panda visit and encourage in visitors of all ages a curiosity about conservation issues.

By June we were about as prepared as we could possibly be. Panda Project Co-ordinator Katie French, an extremely competent, enthusiastic and energetic terrier of a woman, had whipped us all into a black-and-white frenzy. All that was left was for the pandas to arrive.

And arrive they did one cold, winter night. Immediately we realised that nothing, but nothing, could have prepared us for the impact Xiao Xiao, the four-year-old male panda and Fei Fei, the two-and-a-half year old female would have on us, on Taronga and on Sydney generally.

Their two large travelling crates were unloaded at Sydney Airport from the Qantas flight from Melbourne and put into a Panda

Furniture Removalist truck for the police-escorted trip to Taronga. All red traffic lights were switched off and sirens were turned on during the 45-minute journey to Mosman so the VIPs (very important pandas) could have a safe and speedy arrival at the Zoo.

The jostling media were there in force to meet the pandas but because of quarantine regulations and animal welfare considerations, all the photographers got that night were pictures of the wooden crates. Surprisingly, this seemed to somehow satisfy, and this enthusiastic welcome from the media heralded three months of sustained interest when the slightest piece of panda gossip was front-page news. Such is the magnetism of these amazing animals.

Ultimately, with the pandas settled and the official vice-regal welcome ceremony over, Taronga opened its gates to greet the first of the 650,000-strong crowd of panda fans that visited in the next three months.

DID I SAY THE PANDAS were settled? It may have seemed that way initially, but I am sure these innocent-looking, bear-like animals, with their perfect faces, round ears and appealing black eyes used to lie awake at night on their purpose-built futons dreaming up ways to unnerve and outsmart us. The female, Fei Fei, was particularly resourceful and devised many activities just to unsettle us.

We all remember well the time naughty Fei Fei decided to climb out of the exhibit and onto the roof of her night-den. She had only been in our care for about a week and there she was, sitting in the shade of an overhanging Moreton Bay fig tree, gazing across Sydney Harbour not realising, or perhaps very definitely realising, the panic she was causing. She was encouraged down to ground level with bribes of her favourite sugar cane by her extremely anxious keepers. The panda exhibit had to be quickly modified with clear Perspex to prevent her getting a grip on the vertical bars again.

I was asked to put out a news release requesting the donation of three months supply of petroleum jelly. I was slightly embarrassed at having to issue this rather kinky request but it resulted in a donation from the petroleum jelly manufacturer and these large quantities of this slippery substance, once applied to a variety of walls and fences, were very useful in curtailing Fei Fei's arboreal excursions.

PANDA FACTS

CHINA'S GIANT PANDA (Ailuropoda melanoleuca) grows to about 1.9 m tall and weighs up to 125 kg. Usually one, sometimes two near-naked cubs are born after 45 days' gestation. They are only about 15 cm long and weigh a tiny 100 g. The cub is dependent on its mother for 18 months. The Giant Panda is a solitary animal and, despite being instantly recognisable as the worldwide symbol of conservation, its future is anything but secure because of habitat devastation and the continuation of poaching. They have been protected in China since 1939 but their future lies in the protection of habitat corridors as well, so the remaining small and fragmented populations of pandas can move about, find feed and find mates. All this requires local and international support.

It was apparent very early that Fei Fei was the troublemaker of the two. In fact, I don't think she was a panda at all. She was the cheeky reincarnation of an ancient Chinese concubine who spent her days and nights causing mischief and doing as much as possible to make poor Xiao Xiao's life a misery.

Xiao Xiao, we decided, was the reincarnation of an ancient Chinese philosopher who would rather spend his days sitting by the stream doing not much other than eating, sleeping and thinking. Playful Fei Fei invariably had other ideas and pursued Xiao Xiao unmercifully, jumping and rolling on him, nipping him and stealing his bamboo. Xiao Xiao took it all ever so patiently and when he had really had enough he would just simply move away and try to do his thinking in another corner of the exhibit, hoping that something would distract Fei Fei for a while.

Veterinarian Mr Peng and keeper/nutritionist Mr Yang, who accompanied the pandas from China, paid Taronga's panda keepers a great compliment by leaving the husbandry and management of these valuable animals almost entirely in their hands. Senior zookeeper Frank McFayden and his second in charge Elizabeth George, assisted by Andrew Knight, were the custodians of the precious Giant Pandas while they were at Taronga. I am sure that while it was an enormous responsibility it was very definitely a highlight of their career and an opportunity of a keeper's lifetime to care for these secretive and mystical animals.

EACH NIGHT, WHEN TARONGA'S visiting pandas were safely locked into their night-dens and the keepers had gone home, the panda security team took over. These professional guards were based in the keepers' kitchen adjacent to the panda sleeping quarters to keep an eye on the sleeping pandas. The nights were always fairly quiet and every security guard knows that to survive the lonely hours from around midnight to dawn you need a television and a good horror movie or two to keep you company.

One particular night as the chosen movie was coming to a gruesome climax, the security officer, John Inkson, heard some strange sounds coming from the panda bedroom and had a very eerie feeling, almost as though someone was staring at him behind his back.

He jumped up and turned around quickly to see Fei Fei looking through the viewing window in the dividing wall, completely engrossed in the horror movie, too. She had pushed her timber platform bed over to the window, stood on it on tiptoes and with her chin resting on the windowsill was, I am sure, cheering for the villain!

Everyone at Taronga worked extremely long hours during the three months the pandas were in our Zoo. The time just disappeared in a black-and-white blur of early mornings, late nights and weekends spent at the Zoo with media, VIPs and tens of thousands of ordinary panda enthusiasts who all adored Xiao Xiao and Fei Fei as much as we did. Many of the Zoo staff, from utter exhaustion, developed black circles under their eyes and started to look a lot like the pandas themselves.

There was, however, still time for a party and particularly one for Fei Fei's third birthday celebration. We invited children from the local primary schools to come in for an early morning visit. This is always a wonderful time of day in the Zoo and a time when the Giant Pandas were particularly active.

I invited the media and, as was the case with all things panda, they arrived in droves. Unfortunately, this was not one of my most successful media calls.

Lizzie George made a special soy flour birthday cake with bamboo candles for Fei Fei, which we felt sure she would love.

When everyone was ready the keepers opened the birthday girl's night-den door expecting her to wander out and come down the front of the exhibit as she usually did first thing in the morning.

This morning, however, we all waited and waited and waited. The children called over and over: 'Come out, Fei Fei.' We tried singing: 'Happy birthday, dear Fei Fei' very loudly but all to no avail. There was no sign of the precocious panda. Lizzie went behind-the-scenes to see if she could encourage some activity only to find naughty Fei Fei lying on her futon with her eyes shut tight and her paws over her ears. She was simply not in the mood for a party and was making it very obvious. She finally deigned to make an appearance about mid-morning when all her party guests and the media had long gone.

Maybe it was our singing that upset Fei Fei that particular morning because she seemed to like other kinds of musical presentations. As part of the overall festivities we had Chinese musicians wandering the Zoo grounds playing traditional stringed instruments which sound to me a little like an out-of-tune violin. Whenever they were near the panda exhibit and within Fei Fei's earshot, she would stop whatever she was doing, sit back with a dreamy expression on her face and listen to the music. The sounds seemed to transport her to the far-off misty mountains of Guangdong province, her homeland in China. It seemed to calm her and if I were Xiao Xiao I would have paid those wandering musicians to spend a lot more of their time around the panda exhibit. While Fei Fei was busy listening to the music she was also letting Xiao rest in peace.

AS THEIR SHORT VISIT drew to a close, we began to contemplate the unthinkable — a Zoo without Giant Pandas. No more long, snaking queues of visitors, no more black, white and red uniforms, no more yummy Chinese food at the Zoo cafes, no more piles of bamboo, no more calligraphers and wandering minstrels and no more Xiao Xiao and Fei Fei. We had only shared Taronga with these two pandas for three short months, but it seemed like we had known them forever and we would miss them terribly. We had heard about post-panda depression and it was just about to set in at Taronga.

People often say to me what fabulous publicity the Zoo receives, but how can we miss with animals as photogenic as these? The Giant Panda, in all its black-and-white beauty, is made for black-and-white photography and Rick Stevens particularly seemed to specialise in making these animals look even more fabulous.

Finally, the bamboo curtain came down and it was time for the pandas to move on. Xiao Xiao and Fei Fei were crated up once again and taken to Sydney Airport for their flight to Auckland Zoo where they would stay for a further three months to delight and inspire even more people before returning to China.

I accompanied panda co-ordinator Katie French, public relations officer David Alrich, the forlorn group of keepers Frank, Elizabeth and Andrew and sniffling security officers Wayne, Chris and John in a tearful farewell on the tarmac. In New Zealand, the pandas were to be in the safe hands of the Chinese wildlife officers and Auckland Zookeeper Maria Finnigan and her team, but even today I can't stop wondering what became of that amazing pair of animals.

There has been little or no contact with China since then, but we did have an unconfirmed report that poor Xiao Xiao died quite young. I only hope that Fei Fei, because of her tenacity and energy and the substantial funding she helped raise for panda conservation during her international globetrotting, lived a long, happy and productive life back in her misty mountains.

GIANT PANDA KITCHEN

THE ZOO'S BAMBOO HOTLINE, established in advance of the panda visit, registered local stands of bamboo suitable for collection by Taronga's horticulturists. This ensured that Xiao Xiao and Fei Fei had a constant source of fresh and appropriate bamboo. They also received sugar cane as a treat. They arrived with their own recipe for 'Panda Cakes', which were made fresh by the keepers each day. These unappetising-looking creations were made up of polenta, bran, rice flour, soy flour and salt and rolled into orange-sized balls. Every morning and afternoon Xiao Xiao and Fei Fei also enjoyed bowls of warm milk formula consisting of cows' milk and raw egg laced with cod liver oil, vitamin and mineral drops and micro-elements.

They were both exemplary ambassadors for conservation and helped greatly to highlight the need to protect all wildlife and the environment. We were, indeed, fortunate to have had the honour to meet them.

JUST AS DEAR TO MY HEART as the Giant Panda is the Red Panda. This small, brightly coloured cousin of the Giant Panda is also known as the Lesser Panda but I have never thought there is anything 'lesser' about this beautiful russet-red tree dweller from the high altitudes of Nepal and China.

The Red Panda with its bushy tail, exquisite cream markings and toy-like appearance always looks to me as if it should sit on your bed with your pyjamas tucked up inside. I must have had such a creature as a child!

RED PANDA FACTS

THE RED PANDA (*Ailurus fulgens*) is crepuscular, foraging for food at dawn and dusk. The best time to see Taronga's Red Pandas is early in the morning or last thing in the afternoon.

The Red Panda, about the size of a large domestic cat, looks innocent enough but it has extremely sharp claws for climbing and communicates with short whistles and squeaks. It can hiss aggressively and loudly like a cat and also utters short, deep grunts a bit like a bear. A Red Panda scent-marks its territory with droppings, urine and powerful musk-like secretions from the anal glands. The female makes a nest in a bamboo thicket or in a tree-hole. The nest is lined with leaves and other soft plant material. Here, for about three months, she rears her offspring, which can number up to five but is more usually two. Often when there are cubs in the nest at Taronga, the panda keepers set up a 'nestcam' so visitors can see, via a lipstick camera, the tiny cubs developing in the nest box.

Red Pandas first came to the Sydney zoo in the mid-1970s and very quickly commenced a highly successful breeding program which has added more than 30 Taronga-bred Red Pandas to the world's numbers of this threatened species. Taronga now is the regional studbook co-ordinator for this species and organises computer-dating opportunities which enable Red Pandas to travel around the world to spread their precious genes as effectively as possible. Taronga has also established an education program in Nepalese schools to highlight the plight of the Red Panda and encourage community care for this animal. Absolute enthusiast, Carol Bach, is responsible for much of Taronga's great work with Red Panda conservation.

A PARTICULAR FEMALE Red Panda by the name of Meesing — because she was often missing — seemed to have specific and numerous travel plans of her own. These generally didn't take her beyond the wilds of Mosman, but always caused great concern both within Taronga and in the surrounding neighbourhood.

The Red Panda exhibit is situated on the south side of the Zoo and contains lush green undergrowth and tall Norfolk Island pine trees. These most arboreal of animals spend their days exploring the magnificent old trees just made for these expert climbers, enjoying the cool breezes off the harbour and, from a great height, surveying the surrounding countryside. Their partly retractable claws enable them to climb well and in the wild they use trees not only for feeding but also to escape ground-based predators and to sunbathe high in the canopy during the severe winters.

They do come to ground at feed time at the Zoo, and it is wonderful to be able to see these gorgeous animals up close as they munch their bamboo and chomp their way through their choice fruit, vegetables and hard-boiled eggs. It is even more special when there are baby pandas around, which is usually about Easter each year. I always think these perfect little replicas of their parents look like wind-up toys and are almost too beautiful to be true. Red Panda cubs are usually Taronga's first New Year babies, being born in very early January. They are blind and helpless at birth, a bit like a kitten, and are cared for by the mother in a secluded nest box before making their debut about three months later.

The first job of the day for a Red Panda keeper, as for any zookeeper, is to check the animal exhibits to make sure all their charges are present and accounted for and are healthy and happy.

This early morning headcount for the panda keepers frequently involved reporting that Meesing was missing! One particular day, this occurred after heavy rain when the branches of the trees were sodden and perhaps had dipped too close to the fence. It enabled the athletic Meesing to leap onto the enclosure handrail and make her escape. Sometimes she would stay within the Zoo grounds and often make her own way back to her enclosure, usually when the temptation of all that yummy bamboo and those boiled eggs became too much.

Other times she would be missing for days which gave rise to intensive search parties inside and outside the Zoo grounds, an alert to the local police, advice to the quarantine authorities and a request to the nearby national park ranger to be on the lookout for the escapee. There are numerous stands of bamboo around the outside of the Zoo grounds and right throughout the surrounding suburb, so she would never go hungry. Our main worry, however, apart from the fact that we could not possibly lose such a precious animal, was that she may come to the ground and be attacked by suburban pets.

One time the wandering Meesing nearly caused an international incident when she turned up in the garden of the neighbouring army barracks at Middle Head overlooking Sydney Harbour. A soldier, on exchange from Malaysia, was sitting relaxing with a 'quiet ale' on the verandah of the mess, admiring the amazing view from the barracks, the native bushland and the local bird life when he spotted the naughty Meesing sitting in a nearby gum tree looking a bit like a koala in a colourful fancy dress costume.

He almost choked on his beer as he ran for assistance and before she knew it, Meesing had been captured in a true, efficient military manner. A phone call to Taronga confirmed that there had been an escape and very quickly the offender was returned to her 'barracks'.

Another time she did a disappearing act for several days but perhaps having learned from her previous escapade, headed west into suburban Mosman instead of east towards army headquarters. A local Mosman mum was hanging out her

RED PANDA PANTRY

CURRENTLY TARONGA ZOO has three Red Pandas and each day they are served a kilogram of omnivore mix. This is a mixture of apple, pear, kiwifruit or grapes, banana, sweet potato and rockmelon, which is all cut into 2 cm pieces. It is sprinkled with yummy fly pupae and a multi-vitamin powder. Each panda also receives three to six sticks of bamboo depending on how leafy the sticks are.

washing in the backyard one morning when Meesing dropped in to say hello. This backyard was full of bamboo and was just what a wandering Red Panda was looking for. The woman knew immediately that this was definitely not your native species and phoned Taronga. She also phoned the local media and Meesing's exploits made front-page news.

These, and in fact, any photographs of Red Pandas over the years just served to highlight the incredible beauty and appeal of these animals and to reinforce my belief that there is nothing lesser about these pandas.

CHAPTER 9

Elephants I'll Never Forget

History has recorded that wild animals have been kept or been given as gifts by everyone from the Egyptians to European monarchs, but these private royal menageries were not really zoos as we know them now. The word 'zoo' refers strictly to the display of animals for public enjoyment and entertainment and these really have their origins around the mid–1700s.

From these very earliest days, the much-loved and admired elephant has had a high profile in zoos around the world and has played a significant role in encouraging a wonder of nature in visitors.

Taronga Zoo's animal collection has long included both Asiatic Elephants (*Elephas maximas*) and an African Elephant (*Loxodonta africana*). Western Plains Zoo displays the larger African species. As a zoo newcomer, I remembered which elephant species was which by determining that the Asian or Indian Elephant had ears shaped like a map of India while the African Elephant's ears resembled a map of Africa.

IN 1916 JESSIE THE ASIATIC ELEPHANT sailed across Sydney Harbour on a barge laden with a collection of animals from the old zoo at Moore Park to her new home in Sydney's much-heralded and soon-to-be-opened world-class Taronga Zoo at Mosman.

Old Jessie gave millions of rides to eager Zoo visitors, first at Moore Park Zoo and later at Taronga. This famous elephant became a household word when her name entered common Australian parlance: an elderly person was sometimes described as being 'as old as Jessie', someone with a lot of nerve was said to have 'the hide of Jessie' or a rather overweight person could be 'as big as Jessie'.

When the elephant exhibit at Taronga received a substantial upgrade and facelift in 1987 we dedicated these changes, which were beneficial to the elephants and the visitors who came to see them, to the memory of Jessie, a very special elephant and a part of Taronga's amazing and colourful history.

I CAN WELL REMEMBER an elephant ride at Taronga being the highlight of my childhood visits to the Zoo. These visits usually coincided with birthday celebrations. In my youth it would have been Sarina the elephant who was on duty and when I began working at Taronga, Sarina was still very much a large and active part of Zoo life. Sarina, also an Asiatic Elephant, had arrived at Taronga on 4 December 1936 as a baby. Newspaper clippings of

ELEPHANT FACTS

THE LARGEST LIVING ANIMALS on land — the male African Elephant may grow to a height of 4 m and weigh a massive 10 tonnes, while the Asian species is slightly smaller — the elephant is characterised by pillar-like legs, thick-set body and convexly curved spine, large ears and of course, the long mobile trunk. African (*Loxodonta africana*) and Asian Elephants (*Elephas maximus*) live in savannah and light forest. The African Forest Elephant (*Loxodonta cyclotis*), which was formerly regarded as a sub-species of the African Elephant, has recently been given species status. It is a smaller, darker elephant with more rounded ears and a hairier trunk. Its parallel tusks point downwards which enable it to move freely through its more denser forest habitat.

The African Elephant is the largest of the elephant species and can weigh up to 7 tonnes and measure 5 m in length. Both the male and the female of this species possess large forward-curving tusks, which are actually incisor teeth, and are sometimes used as tools in food gathering. African Elephants have two opposing, finger-like processes or outgrowths at the tip of the trunk while Asiatic Elephants have only one. These are used to pick up small objects.

The elephant's life span of about 60 years is the

> longest of any other mammal except humans, and for their whole life they continue to grow.
>
> Elephants live in family groups that consist of the oldest, most experienced female and other females of various ages and their young. Males only join the herd when a female is sexually receptive and otherwise are solitary or, as with young bulls, form bachelor groups.

the day describe her as a 'dwarf' elephant, but she became anything but that. It is estimated that Sarina was born in 1933 and came from Basapa, Singapore.

I told Dave Cody, Taronga's elephant man extraordinaire, that I could remember as a child pulling a strong, bristle-like hair out the elephant's back when I was nervously swaying around on top of what seemed like a grey, four-storey building. I don't know how I ever plucked up the courage to do such a thing. Dave told me that he always knew when some irritating child had pulled a hair out of the elephant as Sarina's usually steady and graceful gait missed a beat and she did a little skip. I am sure that, back then, that would have made me almost faint with fright.

The tradition of elephant rides at Taronga was being phased out when I arrived in 1975. Zoos around the world were struggling with the philosophical question as to whether this somewhat Victorian tradition, which was now seen as somewhat degrading for the elephant, had a place in the operations of a modern zoo.

Taronga Zoo was not only wrestling with that vexed question, but also the fact that the vital flat space used for the elephant ride track could probably be used more effectively. As Taronga is built on the side of a steep slope, flat land is always at a premium and the Zoo Director, Dr Peter Crowcroft, was considering the elephant trail site as a Zoo farmyard. This new interactive exhibit featuring a wide range of domestic animals, cow milking and sheep shearing demonstrations, would give city kids an exciting opportunity to enjoy a taste of country life.

Added to the desire to build this new exhibit was the fact that Sarina was now the only elephant able to give the rides and she would only take her instructions from Dave Cody. Visitors were very disappointed if they came to the Zoo on Dave and Sarina's day off and couldn't have an elephant ride.

So the rides stopped but Sarina was still in great demand at special Zoo events such as Christmas parties where she ushered in Santa in a very grand way. Sarina also earned her keep by testing seat belts with her incredible strength, using her giant feet to test the strength of steel-capped work boots and posing for photographs for Superglue advertisements. I seem to recall she was stronger than the seat belts but the steel-capped boots got the better of her. Maybe she was just being polite by not stamping too hard on the boots. It is still old Sarina's image that is used on the Superglue packets today. She is immortalised on the supermarket shelves!

Dave still took Sarina for regular walks around the Zoo grounds, as much for his enjoyment as for hers, and would even bring her up to the main office where she made a trunk call or two. I can't imagine why Sarina was so well-mannered considering that Dave, her master, or mahout, good friend and the leader of her 'herd', was a very cheeky person with a wicked sense of humour. Perhaps elephants have similar personalities.

Occasionally I needed to work, fortunately quite briefly, on the

Elephants I'll Never Forget

Zoo switchboard while the overworked but extremely patient operator, Helen Barron, took a lunch break. It was a blunderbuss of a thing called a Sylvester switchboard with lots of cords and plugs, and caused me many sleepless nights as I tried to remember which plug went where in anticipation of the dreaded switch duty the next day.

There were so many cords you really needed to be an octopus to operate it, but all of a sudden one day I felt I was at last mastering the beast. Every caller was plugged into the right socket and all was quiet for a blessed moment. Then along came Dave Cody and his elephant.

He got Sarina to put her trunk in the switchboard room window and with one deft sweep the elephant used her dextrous trunk to pull out every cord. The switchboard lit up almost immediately and started ringing off the hook as everyone tried to call back at the same time enquiring why he or she had been disconnected. Not one person believed my excuse that an elephant had put its trunk in the window and had cut off all the calls.

Long after the Sylvester switchboard was nothing more than a museum piece, the telecommunications system at Taronga still seemed to have its problems. There just didn't seem to be a plausible explanation for this, but we were often losing calls, which always seemed to happen to me right in the middle of a radio interview. The excuse I used was that the elephants stepped on the cables. Callers seemed to happily accept this little bit of Zoo folklore.

JILL AND JOAN, both Asiatic Elephants like Sarina, gave me my first experience of animal tragedy at Taronga. Their deaths in 1976 and 1977 had a huge impact on me and the rest of the staff.

Zoo archives show that Joan was born in 1958, probably originating in Indo-China and arrived at Taronga via an animal dealer in Singapore, in November 1962.

In July 1976 Joan was suffering from a gangrenous tail as a result of a bite from another elephant. The tail had to be amputated and to do this Zoo veterinarian, Dr Ted Finnie had to sedate the elephant. The surgery went very smoothly under Ted's skilled hands but as Joan was recovering from the sedation she stumbled and fell to the ground, unable, in her still dazed state, to heave her huge bulk back up onto her feet.

Soldiers from the neighbouring 10th Terminal Regiment at Georges Heights arrived with a large crane and sling to assist with rescue efforts. Long into the cold, wet night, I helped my manager, Barbara Purse, make copious cups of tea and raid the cafeteria for sandwiches and pies for hungry and cold zookeepers, the veterinarian and army personnel, as they tried with all their will and might to raise Joan who was lying on her side.

An elephant's enormous weight, which can be up to five tonnes in this elephant species, on its lungs and heart makes it extremely difficult for the animal to breathe. Finally, Joan was upright and a collective cheer went up. She was, however, still very unsteady on her feet and as she swayed her way across to greet her friend Jill, the other elephant looking on, Joan tripped on a step in her enclosure and fell again. This time, despite another huge rescue attempt, she didn't get up. She seemed not to have the strength or the will left and died where she lay.

It is recorded in Zoo archives that Asian Elephant Jill was born in Rangoon Zoo and transferred to Taronga in 1935. She was the elderly matriarch of the Zoo's elephants. She was suffering from severe arthritis for some time, and for humane reasons, was put down in February 1977, some seven months after Joan's death.

This loss left another huge gap in the Zoo 'family' and a shared sadness quite difficult to describe.

SARINA WAS STILL hail and hearty though and continued to be very much part of a Zoo day. Often she would appear in TV commercials or on daytime television. She also regularly needed and enjoyed a pedicure. Dave would file her toenails with a huge rasp and she would stand patiently on three legs as he filed and filed and then painted her nails with a protective, yellow substance called neats-foot oil.

It always made for a good newspaper or magazine photograph and one particular time I invited Channel 9's 'Midday Show' in to do a live cross while the pedicure was taking place. Live television is always fraught with dangers and difficulties and I should have known that these would be mammoth-sized if an elephant were involved.

The camera was in place, the reporter was waiting with microphone in hand, Dave was wired up with an earpiece so he could hear the questions from the studio host as he was working away on the elephant's feet, and Sarina was waiting patiently for the pedicure to begin. The plan was that a studio segment would finish, a commercial break would run and then they would cross live to Dave and Sarina. Dave would file away at Sarina's less than dainty toenails and answer questions from the reporter and the studio host at the same time.

The producer called 'action' and Sarina obeyed — literally. She opened the floodgates and let go with an elephant piddle which is the biggest wee known to man. Dave was nearly drowned as he was surrounded with gallons of elephant urine, the reporter ran away screaming that her new shoes were ruined and the studio host collapsed into fits of laughter. Dave, well used to such elephant-sized puddles, kept filing and virtually had to carry the segment himself, asking and answering his own questions, while Sarina stood patiently on three legs wondering what all the fuss was about.

THE DEATHS OF JOAN AND JILL left only Sarina living in Taronga's familiar old Elephant Temple exhibit. She enjoyed company so younger, untrained elephants Heman, a bull, and Ranee, a younger female, were moved from an off-exhibit area in the Zoo to this main display area, which underwent some renovation in preparation for their arrival. The surface of the exhibit area was levelled, the surrounding bars were removed, moats were constructed and the Zoo horticulturists planted sweet-smelling gardenia bushes and bright-orange African orchid trees around the perimeter, softening all the edges. Mud and dust baths were installed, too, which are essential for elephant grooming, skin care and protection from insect bites.

Heman the elephant came to Taronga Zoo in 1963 via Singapore. It is estimated that he was born in the wild around 1956. Ranee was born in Cambodia around the same time and came to Taronga, also via an animal dealer in Singapore, a year before Heman. They were both youngsters compared with the now quite elderly Sarina.

The transfer of the elephants to their new enclosure in October 1983 was another job for the army who generously loaned a low-loader truck onto which the two elephants, in a sedated state, were loaded and chained for the short journey in the Zoo grounds.

Heman, normally an aggressive chap, was very relaxed and amenable as a result of the sedative. So relaxed was Heman that his very large and impressive penis was rather embarrassingly hanging down, touching the ground. The transfer process was lengthy, too, and although it began at dawn, the Zoo was well and truly open to visitors by the time the animals arrived at the Elephant Temple.

Of course, a large crowd of interested Zoo visitors of all ages had gathered and we all stood with the invited media watching the keepers carefully unload the elephants into their new home. One little girl asked very loudly: 'Oh, look, Daddy, why has that elephant got two trunks?' Instead of taking this marvellous opportunity to deliver a very well-illustrated lesson on the facts of life, the red-faced dad opted for that universally safe response: 'Ask your mother.'

It was hoped that Heman could be housed together with Sarina and Ranee but he either beat up the females or they ganged up on him and gave him a hard time, so eventually the three had to be separated. The females, however, got on very well together in these new living arrangements.

CHORI, THE ONLY AFRICAN ELEPHANT in Australia at the time, also lived at Taronga Zoo in my early days there. Chori had been wild-caught in Africa as a calf in the early 1940s and was to be a gift for Hitler, but her life was changed forever with the outbreak of World War II. The ship she was sailing on from her homeland to Europe was diverted and she ended up in Sydney and lived her whole life at Taronga.

Reputation had it that Chori was a cranky elephant. She would take a dislike to certain Zoo staff for no apparent reason and would throw small rocks and pebbles at them as they passed by.

Chori had the most amazing gnarled, cracked, grey skin. She had beautiful brown eyes with long, long eyelashes. There always seemed to be a trickle of a tear from the corner of these eyes. I know her keepers did as much as they could to keep her occupied but it never seemed enough for this lone elephant.

After Western Plains Zoo at Dubbo was established, the Zoo Board decided it would be far better for Chori, now in the twilight of her years, to enjoy the larger spaces of the open-range Zoo for the time she had left. It was late 1978, from memory, and Chori had been at Taronga since the mid–1940s.

Moving an elephant is always a complex and risky task, especially an untrained and unmanageable animal such as Chori, but everyone wanted to have a go to give her the chance of a different life. She needed to be sedated enough so Dave and the other keepers could approach and get chains on her and then guide her through her night house into a crate for the road journey west.

We had planned a celebration party that night once Chori was safely boxed. Everyone, especially Dave Cody, Graham Button and David Butcher, the veterinarian and first officer-in-charge of Western Plains Zoo, wanted to wish the old elephant a fond farewell and safe journey to her new home out west.

Sadly, we all met after work but it was not for a celebration. It was more like a wake. Chori had slipped in the night house and had fallen to the ground, unable to be lifted up in the confined space. She never made it to Western Plains Zoo. This was yet another sad ending to the tale of an elephant at Taronga Zoo. Taronga seemed to be filled with crying people that night.

I still have a bracelet made by Dave from one of her tail hairs, which he gave me as a reminder of old Chori.

FOR MANY YEARS, as Sarina walked around the beautiful grounds of Taronga, she would stop at a kiosk near the Elephant Temple, put her trunk in the roller shutter window and receive a sticky bun or two from the women who worked there. She used to get quite cranky if the kiosk was closed and Dave had some trouble getting her to move on without her usual treat. I should note here that feeding is an elephant's principal activity. It digests only about 40 per cent of its food and in the wild an adult elephant would spend up to 16 hours a day searching for and eating the vast amounts necessary to fuel its enormous body. An elephant uses its extremely dextrous trunk, tusks, head and body strength in this feeding process. I am sure the kiosk staff were aware of these facts and tried to be as quick as possible handing over the buns in case Sarina became a little too demanding.

Sarina reached the age of 57 before her health began to deteriorate. She had a heart attack in 1979 and was in such pain and discomfort that the veterinarian, Ted Finnie, and Dave made that very difficult but most humane of decisions to put her down. How Dave could make that decision I'll never know. Sarina, now lying on her side and in great distress, still greeted him that fateful morning with her usual warm and friendly rumble. Yet again, I shared tears with my mate Dave.

We calculated that during her 33 years at Taronga Sarina would have given rides to about 1,700,000 delighted visitors. Dave says she was the most intelligent and clever elephant he ever had the good fortune to work with.

When the old kiosk next door to the Elephant Temple was demolished and a new coffee shop was opened in 1988, I suggested it be named 'Sarina's' in memory of the elephant with a very sweet tooth. She would have loved all the yummy cakes, scones and pastries now available at the cafe.

THE NEW FACILITIES for African Elephants at Western Plains Zoo were completed in 1978 and Dave Cody had gone to the United Kingdom the year before to bring back four young elephants from Whipsnade Zoo in England. The young male elephant, David, and three equally youthful females, Cherie, YumYum and Cuddles travelled approximately 20,000 kilometres by sea with Dave. The long voyage took around six weeks but Dave and the animals were good sailors and arrived in Sydney hale and hearty. The elephants stayed for about six months in quarantine at Taronga and then travelled by road to Dubbo. Dave went with them to ensure their safe arrival at Western Plains Zoo.

David, the young male, was never a very strong animal. He had rheumatoid arthritis in his back legs, which is not uncommon in captive African Elephants. Sadly he did not live past the age of ten. Cherie, YumYum and Cuddles, however, grew into healthy strong adults and are still very much the centre of attention at Western Plains Zoo.

In 1983 an opportunity to acquire two more African Elephants from a wildlife park in the United States saw Assistant Head Keeper, Tony Carrick, bring huge bull elephant Congo and female Toto, by sea on a container ship from the United States. The animals travelled in a furniture pantechnicon on the deck and for 17 days Tony fed and shovelled as Congo and Toto swayed their way across the Pacific Ocean.

The day they arrived in Sydney the weather forecast was for heat wave conditions. The ship docked at the White Bay container terminal where we had a low-loader truck waiting to take the pantechnicon and police motorbike escort to get us out of the city precincts as quickly as possible.

The unloading operation started at 7.00 a.m. and already it was 30 degrees Celsius. The arrival of these elephants was an interesting news story as Congo was described in America as the 'world's largest elephant' and a movie star of some note, so there was a strong media contingent with us at the dock to watch the unloading.

The Department of Quarantine permitted the elephants to go directly to Western Plains Zoo instead of having to spend time at Taronga, so we set off on the long road journey to Dubbo with Dave Cody travelling in the cabin of the truck taking the elephants. Tony Carrick, veterinarian Lesley Reddacliff and myself followed in a Zoo car to keep an eye on the truck from behind.

As we travelled slowly west on this hazy February day the mercury began to rise and when we pulled up for a break at Lithgow, just west of the Blue Mountains, it was obvious that the elephants were feeling the heat. Dave Cody climbed onto the roof of the pantechnicon and hacked some holes in the front and back to let air in to cool the animals. He also opened a small door on the kerb side of the pantechnicon, which meant that Congo's trunk was very visible for the rest of the trip as he investigated every aspect of the unusual situation he had got himself into.

We stopped again about half an hour later in a little hamlet on the way to Bathurst and by this time it was 38 degrees. The elephants needed a drink of water badly and we had run out of supplies. I was elected to go into the closest house and ask if we could have some water from the garden hose. I knocked on the front door and asked my innocent question: 'May I trouble you for some water for our two elephants, please?' When the woman's face registered incredible disbelief I realised how absurd the request must have sounded, but after more explanation she kindly enabled us to hose down the grateful elephants and give them a huge drink from her precious tank water.

We needed to repeat this about every hour so that the road trip to Dubbo, which usually takes about five and a half hours, was really dragging on and it looked like we might not arrive until dark.

I don't know whether it was the heat, or the slowness of the journey, or the thought of the cold beer that was beckoning in Dubbo, but the truck driver suddenly began to put his foot down. As the truck lumbered into the twilight, bits of rubber started flying off the tyres. The pantechnicon with its precious cargo was swaying and leaning at every corner and Tony, Lesley and I started to get very nervous watching from some distance behind. Tony got so nervous he lit up a cigarette, which was the first one he had ever smoked. His explanation for starting to smoke was: 'I haven't brought the elephants all the way from America to lose them at bloody Molong!'

We switched off our little car's air conditioning to give us more power and finally overtook the hurtling truck to demand that the driver slow down. Both Dave and the driver were completely unaware of how dangerous it had looked from behind and seemed quite amazed at our distress.

We all cooled down, with a hose for the elephants and yet another soft drink for the perspiring humans and finally made it into Western Plains Zoo around 7.00 p.m. This was 12 long, hot and stressful hours after leaving Sydney. The temperature in Dubbo was still 40 degrees.

Congo and Toto were immediately put into the elephant yard adjacent to resident elephants Cherie, YumYum and Cuddles who could not believe their eyes when the two new arrivals made their first appearance from the back of the truck.

Those shameless female elephants ran trumpeting across their paddock, great ears flapping and trunks slapping the ground and snorting with glee. They were so excited to see Congo they even turned around and shook their substantial backsides in his direction. I'll always remember this outrageous pachyderm welcome for the newcomers. I don't think the female elephants were very interested in Toto's arrival at all. It was definitely her very large and impressive friend, Congo, who the girls were pleased to see.

Toto must have been jealous of all this attention Congo was receiving because instead of being happy and relieved to be on dry land once more, she worked herself up into a very cranky mood indeed. For quarantine purposes, the veterinarian needed to spray her feet with disinfectant before she could set foot on Australian soil. She took an instant dislike to this procedure and once out of the truck, chased poor Lesley Reddacliff into the moat just to show everyone, including those onlookers in the next exhibit, that she was not to be messed with.

When I returned to Taronga I put in a petty cash claim for 53 bottles of juice and soft drinks. I think the Zoo's business manager thought I'd been giving them to the elephants. I had a hard time convincing him the drinks had all been consumed by four hot and thirsty Zoo staff on that journey west.

Congo and Toto settled in and the African Elephant herd of five at Western Plains Zoo looked very impressive. Unfortunately no elephant births have ever occurred here or at any zoo in Australia.

There was a mouse plague in Western New South Wales in October 1984, which caused numerous problems for the Zoo. As a result, poor Toto died of a virus infection carried by the mice. Congo, always such an impressive elephant, lived until the year 2000 when he died suddenly of pneumonia.

In December 1982 Burma, an Asiatic Elephant from Bullens Circus was offered to Taronga Zoo. Burma had come to Australia via Thailand where it is estimated she was born in 1952. She was described as a 'problem elephant' as she had knocked down her trainer. Her bad behaviour had meant she was not suitable for circus life any more and poor Burma had been chained to the spot, alone in a paddock for some years. She found a new home at Taronga and moved in with Ranee in the 'Hollywood Moorish'-style Elephant Temple overlooking the harbour.

Heman, meanwhile, lived in the adjacent house and yard. The Elephant Temple is close to the Zoo's outdoor concert stage and I used to wonder if Burma ever recalled her circus days as the music wafted up from the many beautiful concerts held there. She often swayed from side to side but that was probably more from having been tethered at the circus for long periods than from any kind of music appreciation.

Ranee developed vulval papilloma that became very large, obvious, quite ugly, and very uncomfortable for her. It really worried us that she was often the subject of ridicule as visitors pointed to her quite disfigured backside. Animals in zoos should be admired, not laughed at.

It is a big decision to knock out an elephant and the operation required to remove the large protuberances was a complex procedure. It had to be done and in November 1999 the Zoo veterinary team gathered assistance and advice from all round the country in preparation for the surgery. It was like a well-planned military manoeuvre and the lengthy and complicated operation ran exceptionally smoothly. Ranee was back up on her feet and everything was looking fine until she started showing signs of abdominal bloat. Despite exhaustive efforts by the vets and her keepers day and night, Ranee died four days after the operation. She was 43 years old. I have to say I was somewhat comforted by the fact that Ranee was buried alongside the Public Relations office. It was somehow reassuring to have her there and to see the elephant ears the gardeners had planted on her huge grave wave in the wind. It was almost as if she was still around.

Elephants I'll Never Forget

Now only two elephants remain at Taronga Zoo gazing at their harbour views. Heman and Burma are hosed with strong water sprays every day. I also hasten to add that if the public relations manager was ever walking by while keeper Ian Chisholm was in charge of the hose then she, too, could expect a hose down! Considerable time is taken to hide cubes, carrots and nuts under logs, in tractor tyres and in tree trunks to encourage the intelligent elephants to search for their food. They have swimming pools, mud and dust baths and are also provided with substantial amounts of browse material. I always loved watching the elephants strip banana palms in one sweep, deftly break Moreton Bay fig branches with their strong teeth and use bamboo sticks like giant toothpicks. Heman often likes to 'wear' his breakfast on his head and can be seen sporting a jaunty 'beret' of hay that he seems to keep for morning tea.

HEMAN AND RANEE, like many of us, enjoyed a day they will never forget on Friday 15 September 2000. It was the day the Olympic Torch Relay passed through the grounds of Taronga Zoo on the final leg of its long journey to Homebush Bay in time for the opening ceremony of the Sydney Olympic Games.

Ten thousand people gathered at Taronga that glorious spring day to welcome the Torch and the runners, and to be touched by the spirit of the Olympics. The Torch arrived by ferry at the Zoo wharf, soared high above Taronga on the Sky Safari to the top of the Zoo and then wound its way down through Sydney's beautiful zoological gardens past the Australian animals, the famous Floral Clock, the Free Flight Bird Show, Seal Bay and, of course, the elephants.

I am very fortunate that this day, one of my last special ones at Taronga, has given me such wonderful memories. I can still see

Heman and Burma, with Sydney shining in the background, watching as Torch Relay runner, Olympic yachtsman John Bertrand, posed for photographs in front of these two dear old Taronga identities.

I began this chapter describing the philosophical quandary Taronga faced in the mid–1970s deciding whether or not elephant rides have a place in the modern zoo. I now find myself, having recalled my somewhat sad history of elephants at our two Zoos, questioning whether there is a place for elephants at all in zoos in the 21st century.

I should note here that naturally, over the past 25 years, husbandry techniques, veterinary management and exhibit design for elephants have evolved and progressed for the benefit of the elephants and all zoo animals.

It is difficult, however, to imagine a zoo without elephants. I am particularly conscious of this whenever I visit with my young nephew, Jack, as it is the huge elephants with their swaying trunks that have the most positive impact on this five-year-old, and thousands of others.

Taronga is currently making extensive plans for great changes to the way these gentle grey giants are housed and displayed in Sydney's zoo. Conceptual plans show an Asian habitat, with the elephants as the centrepiece and surrounded by other animals of the region such as binturongs, tapirs, otters, gibbons and langurs. A winding pathway will take visitors through different environments, all the while catching glimpses of young elephants in a clearing, working actively as they would in the jungles of Thailand or India.

As the Zoo contemplates these changes and the possibility of acquiring these new young elephants, my hope is that present-day Zoo Director, Guy Cooper will spare a thought for the elephants I'll never forget.

TRUNK CALLS

TWICE AROUND AN elephant's foot is its height from the ground to shoulder.

Every week at Taronga Zoo Heman and Burma eat 140 kg carrots, two 20 kg bags dairy cubes, 20 kg corn on the cob and 55 bales of lucerne hay. They also enjoy large quantities of Moreton Bay fig and Mulberry Tree branches and bamboo.

The Asiatic Elephant is smaller and lighter than the African Elephant and has one lip on its trunk whereas the African Elephant has two lips.

In Asia, the elephant has long been held in reverence. In fact, the Hindu god of wisdom and prudence, Ganesha, has the head of an elephant.

In the wild, elephants rest in the shade for about four hours during the hottest time of the day. An elephant's large fan-shaped ears contain a network of blood vessels. The ears are constantly in motion to aid heat loss. At Taronga on hot days the elephants are given access to their elephant temple night house for shade. Heman invariably seems to know when the UV level is especially high and disappears inside.

The elephant is an endangered species. The Asian Elephant is endangered due to competition with people for land while the African Elephant populations have been decimated because of hunting for their ivory.

CHAPTER 10

All Creatures Great and Tall

TARONGA'S FOUNDERS, way back in 1912, had great foresight when choosing the site for Sydney's new zoo. The dedication of this piece of Crown Land in Ashton Park at Mosman as a zoological garden meant that Sydney has, forever, the best-located zoo in the world. On the northern shores of magnificent Sydney harbour and only a 12-minute ferry ride from the bustling city, Taronga has stood proudly for 85 years as an icon of Sydney and a highly regarded member of the international zoological community.

Even the name 'Tarong' (later to become the more lilting 'Taronga'), believed to be an Aboriginal word for 'water view' was a perfect choice. There are water views from dozens of vantage and picnic points around the Zoo grounds and many of Taronga's animals, particularly the chimpanzees, Red Pandas, Andean Condors, Asiatic Elephants, orang-utans and Himalayan Tahrs gaze out on this multi-million-dollar vista.

The choice by Zoo Director Ron Strahan in the late 1960s of the paradoxical platypus as Taronga's logo is also very fitting. This strange, aquatic, 'jigsaw' animal that seems, to me, to be designed using parts from many animals, perfectly represents a zoo with a water view.

Taronga's visitors, too, soak up the panoramic views and I know, from personal experience, that those fortunate enough to work there never, ever become blasé about the Zoo's truly wondrous location.

TARONGA'S GIRAFFES (*Giraffa camelopardalis*) have one of the very best views from the Zoo. These elegant giants — the males can grow up to 5.8 metres tall — don't need to strain their necks to look at the harbour activity and at the sparkling city beyond. Photographs of the giraffes with the harbour views as a backdrop have long been synonymous with a visit to Taronga. I often wonder just how many times that particular shot has been taken home as a souvenir snap of a holiday visit to Sydney's world famous Zoo. The giraffes don't know it but they must be in billions of family photograph albums around the world.

This incongruous but quite intoxicating combination of the African Giraffes against the backdrop of Sydney and the harbour has for 85 years attracted many, many Zoo-visiting celebrities. In my time there I had the honour to welcome hundreds of special visitors, from the charismatic Nelson Mandela to the delightful Kidman-Cruise family in happier times, and all manner of celebrity-spotting photographers as well. Two such visits that really stand out were the ones by the

Australian Wallabies Rugby team and the American singer Michael Jackson.

The Wallabies were about to depart for South Africa for a World Cup tournament and Channel 10, the official broadcaster of the series, brought the players to Taronga for media photo opportunities with the giraffes. The sports media turned out in force and I remember thinking that captain John Eales was about as tall as a giraffe himself and equally as dignified.

All team members were resplendent in their new Wallaby uniforms but one particular new player was the centre of the photographers' attention. They wanted him to 'feed the giraffes'. Now, carrots are the only things that keep the giraffes interested and in the right place for the photographs for any length of time, and there is nothing that makes a giraffe salivate and dribble more than a nice juicy carrot (or 20).

Camera shutters were clicking and whirring but, instead of smiling, the young rugby star was starting to look very uncomfortable. That's when I noticed that his pristine Wallaby blazer had giraffe dribble all over the shoulder and down the sleeve and the player was getting extremely upset about the mess. I was just deciding that I had never actually heard of anyone dying of giraffe dribble when I heard Reuters photographer Mark Baker, a New Zealand rugby supporter from birth, mutter, 'If he's upset now, wait until an All Black dribbles on him!'

SINGER MICHAEL JACKSON, accompanied by his substantial entourage, has been to Taronga two or three times during his visits to Sydney. He always visits privately after Zoo hours and brings with him a security force that is the size of a small army. He shows an extensive knowledge of and a deep respect for all animals. During his last visit we took the singer and his party to the giraffes and had carrots at the ready for the obligatory feed and photo opportunity. Michael declined the offer to get close to the giraffes (maybe he had heard about the dribble-damaged blazer incident) saying he was allergic to these animals but he encouraged his guests to enjoy the unique experience. While they were feeding the giraffes I noticed the singer had lifted his signature facemask, and was enjoying munching on one of the giraffes' juicy but unwashed and certainly unpeeled carrots!

Whatever the season, the weather or the time of day, the view at the giraffe exhibit at Taronga is magic. If you walk past early in the morning before the Zoo gates open, the giraffes are peeping out of their night houses watching the keepers sweep and clean their yard as the sun comes up across the harbour. If the rain is pouring down the giraffes take shelter in these very tall buildings, craning their long necks and poking out their amazingly long tongues to collect the raindrops.

As described in Michael Allin's fascinating book *Zarafa,* the word 'giraffe' is derived from the Arabic *zerafa,* a phonetic variant of *zarafa*, which means 'charming' or 'lovely one'. It is a most apt description of this intriguing, curiously graceful animal, which was often presented as a gift of friendship and peace, and has inspired folklore, poems, literature, songs and even fashion.

Giraffes have a lovely, warm, 'horsey' aroma, which is even more obvious if you visit their part of the Zoo at night when your sense of smell is at its sharpest. To see the tall silhouettes of these unhurried, beautiful animals outlined against the twinkling lights of the city is one of the special privileges of a quiet walk through the grounds of Taronga at night. This amazing twilight scene became

Having the Bouquet and Eating it Too

WHEN TARONGA BIRD keeper Liz Notley and marine mammals trainer Steve Romer were married in 1993, they held their wedding reception at the beautiful Taronga Centre in the Zoo grounds overlooking the harbour. Liz and Steve, both dedicated animal people, wanted their wedding day to be very 'zooey' so decided to have their wedding photographs taken with the giraffes as the backdrop. I invited a freelance news photographer to cover this unique wedding day opportunity. Liz's beautiful bridal bouquet contained some sprigs of eucalypt and the giraffes thought that this floral arrangement was made especially for them. Quick as a flash, a very long tongue hooked itself around the tasty gum leaves and a tug of war between the bride and a giraffe followed as poor Liz tried to rescue the remnants of her bouquet. These very funny photographs were seen in magazines around the world.

the signature image of Taronga's NightZoo when it was launched in the spring of 1995, giving visitors extended opportunities to enjoy the elegant giraffes and the spectacular harbour views.

* * *

TARONGA'S GIRAFFES have bred successfully for many years with more than 90 calves being born since the Zoo opened in 1916. These Taronga-bred giraffes have been sent to zoos throughout Australia and overseas. I always found there was something unique about the atmosphere in Taronga when giraffe calves made their debut. The entire Zoo seemed to take on a special feeling. These gangly, two-metre-tall babies with their velvet noses, impossibly long eyelashes surrounding liquid-brown eyes, and with strange 'aeroplane' ears looked, to me, like escapees from the Zoo souvenir shop.

The media loved these photo calls and the announcement of a giraffe calf debut invariably attracted a very large crowd of eager television news crews and photographers. They were always rewarded with exquisite pictures of the proud mum and tiny calf being encouraged from the 'maternity ward' and welcomed to the display yard by the inquisitive but ever so gentle giraffe relatives.

There have been one or two occasions when a giraffe mum at Taronga has either run out of milk or failed to feed her newborn calf properly. Enthusiastic zookeepers like Jane Burgess, Hayden Turner, Anthony Dorrian and Glen Sullivan have stepped in to became surrogate mums to the two-metre baby. I loved watching these dedicated keepers standing on tiptoes to bottle-feed the calf and then use a damp sponge to groom the baby's face and neck, just like the mother would do with her rough tongue. After each four-hourly feed, the keepers would also spend time brushing the animal from head to hoof and this early bonding with humans ultimately made the calf so trusting and quiet — it was like a gangly pet pony that followed its human mum everywhere.

Once these calves were weaned, usually at about 12 months of age, and were tall and strong enough to look after themselves, they were introduced to the rest of the herd. These very tall and curious relatives accepted the strangers politely and unconditionally even though, I'm sure, the newcomers didn't have all the usual and necessary giraffe manners and behaviour.

Many a beer was won and lost by zookeepers making bets while endeavouring to predict when a giraffe would give birth. Somewhere towards the end of the animal's 15-month gestation period, I used to hear comments such as: 'Her udder is full, it won't be long now', but days, sometimes weeks would go by before the calf finally arrived. Often the birth would occur in the middle of the night and the keepers would arrive at work early in the morning to find their giraffe herd had a delightful new addition.

It is a much more exact zoo science today with Animal Watch and data collection programs recording matings and determining subsequent births with almost pinpoint accuracy. An artificial insemination program for the giraffes is being developed at Taronga that will enable such events to be determined with even more certainty — but certainly a lot less fun.

A giraffe calf comes into this world in a rather amazing way because the mother gives birth standing up. The calf, usually a single animal (but twins have been recorded), in the birth sac falls almost two metres to the ground. The baby instinctively manages to turn its body in this brief period from birth to landing, so it avoids ending up on its head. It is then quickly cleaned up by the mum and encouraged to its feet with a sense of urgency. In the wilds of Africa's dry savannahs and open woodlands, the mother giraffe and her new calf, separated from the safety of the herd for the birth process, shelter in the acacia thornbushes. Even so, they are easy prey for lions and leopards and possibly even hyenas. It is imperative, therefore, that the calf gets to its feet as quickly as possible, which is usually after only 30 minutes. The mother and the calf stay slightly separated from the herd for about the first ten days.

> ## GIRAFFE EARLY WARNING SYSTEM
>
> GIRAFFES HAVE THE largest eyes of all land animals. Their exceptional eyesight enables them to identify and communicate with other giraffes visually from as far away as 1.5 km, far beyond scent and sound. They seem to have a unique early warning system that inspired the ancient Egyptians to use the figure of the giraffe in their hieroglyphs to mean 'foretell'.

Photographs of the giraffe birthing process are quite rare but Rick Stevens expressed his interest one time at having a go at recording the imminent event for the *Sydney Morning Herald*. I needed to obtain special permission for this project so that as the estimated confinement date drew near, the expectant mum could be shut into one of the night houses where a makeshift 'hide' had been specially set up for Rick.

It was September 1987 and Sydney was having an exceptionally cool spring. Rick came to the Zoo night after chilly night to crouch in this small draughty space and peer through a hole in the plywood sheeting, while trying to keep very still and quiet so as to not startle the mother-to-be. I think his chattering teeth were his biggest problem.

After about a week of this repeated nocturnal activity by Rick and no activity at all by the pregnant giraffe, the very tired photographer temporarily pulled the pin on the idea. He decided to stay home in his warm bed for a night or two and maybe return to the Zoo for a further attempt later in the week. Of course, the calf was born that very night. I am sure the mother, Faye, was just waiting for a little privacy.

Feeling sorry for the very disappointed Rick, we compensated him a little by naming the healthy male calf 'Ricky'. Ricky, the giraffe not the photographer, subsequently went to live in Auckland Zoo in New Zealand to sire his own calves there.

While the giraffe herd has stood proud and tall in Taronga for a very long time, the look of their exhibit has changed considerably over the years. All the original high chain-wire mesh-fencing that I can remember from my early Zoo years has gone and instead moats now surround the display yard. The Zoo horticulturalists have planted acacia trees and thornbush shrubs and the thatched shade rondavels give the whole area a realistic African atmosphere. These changes also provided a far better view of the animals and the harbour.

WHEN THE ZOOLOGICAL PARKS Board of New South Wales established Western Plains Zoo at Dubbo, the world's tallest animal, the giraffe, head and shoulders above the rest of the animal kingdom, was chosen as the symbol for this proud, new open-range zoo.

The original giraffes sent to Western Plains Zoo all came from Taronga. Just imagine what these animals, born on the side of a hill at Mosman, must have thought when they were loaded up and trucked out to the Western Plains of New South Wales.

The early days of this fabulous new zoo, the first public zoo to be built in New South Wales for over 60 years, were exciting times for the Taronga staff and the animals involved in this innovative project. Gradually, as the planning was completed, the roads and infrastructures built and the enclosures finalised, the animals began moving into the 'Ark' out west, two by two. Not a week went by that we were not either receiving animals for quarantine in transit to Dubbo, crating animals for transfer out west, distributing news releases heralding the arrival of African Elephants, cheetahs or White Rhinoceros or packing to the roof any Taronga truck heading to Dubbo with everything from stocks of stationery to shiny new shovels and supplies of souvenirs!

One of the funniest transport inventories I read at this time listed: eight boxes toilet rolls, seven boxes Zoo letterhead, three boxes Zoo envelopes, 12 bales of lucerne hay, 10 ticket rolls, one adding machine, two Sydney telephone books, one typist's chair and one Giant Anteater.

Crating a giraffe for transport is an exacting task requiring an early start, lots of patience, a large truck, a large crate and an even larger crane. The usually obliging giraffe was enticed into the crate using juicy carrots as a lure. This process was repeated over many weeks, enabling the animal to feel comfortable in the crate before it was ready for transport. The wooden crate was tall, so the giraffe's head was just visible, and sometimes had a detachable shade cover over the top. The animal was usually crated in the late afternoon and remained in the giraffe yard until dawn the next day when the crane arrived to lift it out onto a truck and low-loader. Once loaded the truck (with zookeeper) left Sydney in the cool of the morning, which was the most comfortable time to travel and also avoided the peak-hour traffic out of the city.

I often went separately in a Zoo car, phoning on ahead to regional newspapers and radio stations alerting them to the fact that a Giraffe was about to trundle through their town.

Giraffe Facts

THE GIRAFFE HAS seven vertebrae, which is the same number as a human has.

It also has very keen eyesight and can spot predators from a great distance. The relatively short, powerful back legs mean it can take off in a rush, and an adult can one-kick a lion to death and out-accelerate a horse.

The giraffe is often thought to be mute, but it has been heard to infrequently utter deep guttural bleats. A giraffe's tough, prehensile lips and long narrow tongue, which is coloured blue to protect it from sunburn, enable it to feed on the thorniest bits of an acacia tree.

The regional media turned out in strength, as did wide-eyed school children; even hospital patients, who had been listening to the radio, waved from their wheelchair vantage points on verandahs as the giraffe convoy went by. The interest level in the new zoo was enormous and we always received the warmest of welcomes in the tiny towns and larger cities on the road stretching from the Blue Mountains west to Dubbo, every bend of which we all came to know extremely well.

One particular giraffe must have been taller than the others or maybe we had hired a taller truck! It presented a major problem at the first overhead railway bridge we encountered along the route out of Sydney. Usually the giraffe's head could be bowed sufficiently to negotiate the bridge by lowering the canopy slightly. This time, however, air had to be let out of all 18 truck tyres before we could squeeze under. We then had to limp to the nearest garage to pump ourselves up again before we could continue the journey.

The wonderful new Western Plains Zoo at Dubbo was officially opened by the highly regarded Governor of New South Wales, Sir Roden Cutler, on 28 February 1977. This was a very proud day for all involved.

The westward migratory pattern to Dubbo, very frequent in those early years, still continues today, albeit to a lesser degree. Western Plains Zoo has flourished to become this country's leading open-range zoo with a fine, worldwide reputation for its animal husbandry and exhibiting techniques, its education endeavours and its endangered species conservation programs, particularly for Australian native animals. It's also a marvellous place for relaxation and enjoyment for human animals. Visitors can even stay overnight in luxury tented accommodation and enjoy an African safari in the middle of the Australian bush.

Western Plains Zoo is a true jewel in the zoological world's crown and a credit to its pioneers and its present-day staff. Long may this oasis, on the Western Plains of New South Wales, continue to do great things for environmental education and wildlife conservation.

I CAN'T SAY that all the animals we transported out west were as well behaved as the dignified giraffes. I think that transporting a zebra has to be a zookeeper's worst nightmare. The Zoo usually displays Chapman's Zebra (*Equus burchelli antiquorum*), and I ask you to just think of the worst-behaved horse you have ever known and multiply that by ten — that's any zebra! They are cranky, rearing, biting, kicking machines but I guess when you look that good you don't have to be well mannered as well. These most handsome of creatures which can weigh up to 385 kilograms and grow to a length of 2.5 metres, despite being very poor travellers, always arrived safely in Dubbo but I can't say the same for the accompanying truck driver and keeper. They usually had excruciating headaches from listening to the cranky zebra kick the crate with its strong hooves for the entire six-to seven-hour journey.

I don't know if these keepers would agree with me but I think I could forgive a zebra anything. As far as I am concerned the zebra is 'design perfection'. Its black-and-white markings camouflage it so effectively in the heat haze on the African plains, and its bristling mane and striped stockinged legs make it a most striking and appealing creature and certainly one of the most admired animals in a zoo collection.

ANOTHER HOOFED ANIMAL almost as cranky as the zebra but not nearly as exotic looking is the Mongolian Wild Horse. One of the most significant contributions to wildlife conservation by the Zoological Parks Board of New South Wales was made in 1995 when some of these Mongolian or Przewalski's Horses (*Equus przewalskii*), also known as Takhi, which were bred at Western Plains Zoo, were transferred to Mongolia for release.

In 1982 Western Plains Zoo imported 13 of these rare horses from a wildlife park in England. These primitive-looking animals, measuring up to 2.6 metres in length and weighing up to 300 kilograms, the ancestors of the modern horse and the subject of prehistoric cave paintings, were extinct in the wild until an international zoo effort gave the species a future.

The imported Przewalski's Horse herd flourished in Dubbo and ambitious plans were developed, in concert with several European zoos, to send Western Plains Zoo-bred Takhi, along with stock bred at Monarto Zoo in South Australia, to the steppes of Mongolia for reintroduction to the wild. These animals live in cohesive, long-term herds that wander great distances for grass, leaves and buds. A typical herd is led by a senior mare and has up to four other mares, their offspring and one stallion. This stallion usually lives out on the periphery of the herd. The Takhi are short and stocky compared with domestic horses.

James Woodford the *Sydney Morning Herald*'s environment writer and photographer Rick Stevens put their very enthusiastic hands up to accompany Zoo staff and the horses on this amazing journey to this most desolate part of the world. It was to be quite a remarkable story, and James and Rick did it proud.

Departure date for the shipment was World Environment Day, 5 June 1995. We arranged an official ministerial farewell in Dubbo the day before, then the horses were boxed at dawn the following morning in readiness for the road trip to Sydney International Airport. They were to then board a cargo flight that same evening to China before changing planes for the final leg of the long journey to the wilds of Mongolia.

It was the morning of departure, as we were watching the excited horses being darted and boxed, when Rick decided he might need reading glasses!

I remember thinking what a time to make such a discovery. Here is a man who lives and works by his eyes, who is about to set off on a journey to the other side of the world to live in primitive yurts, load film by torchlight or moonlight to exclusively record every moment of this epic undertaking, and he asks me if the Zoo souvenir shop sells reading glasses! Fortunately it did, albeit the 'one size fits all and suits no one kind', but at least Rick, looking a bit like Dustin Hoffman in *Tootsie*, went off to take his fabulous photographs so we could all eventually share the moment that the Przewalski's Horses, with the Aussie accents, first galloped in the dust on the plains of Mongolia.

WESTERN PLAINS ZOO provided wide, open spaces, deep lakes and rivers at Dubbo for many large animal species previously displayed in much smaller areas at Taronga Zoo. I remember counting at one time 29 old wire and concrete exhibits that were demolished at Taronga to make way for new exhibits such as Chimpanzee Park and the Orang-utan Rainforest. I know I keep saying it but it's this constant regeneration of Sydney's zoo that I, and everyone who works there, always found so stimulating.

One of the largest animals to benefit from these changed conditions was the Common or Nile Hippopotamus (*Hippopotamus amphibius*). These hippos at Taronga lived in small concrete yards, with bathtub-sized ponds, exactly where the orang-utans now live. It must have suited them quite well at the time, as I recall several hippo calves being born in the first few years I was at Taronga. I also remember being embarrassed at my obvious lack of zoological knowledge when I saw my first hippo calf in Taronga in 1977. I innocently asked Head Keeper Graham Button whether the calf, which had been born that particular morning, was a male or a female. Graham, with decades of zoo

Hippopotamus Facts

THE HIPPO IS A truly amphibious mammal by means of valves and press their ears close to their heads to prevent water getting in. Hippopotamus calves, born after a gestation of about 240 days (which is quite short for such a large animal), can swim before they can walk. Because they suckle underwater often, a calf is immediately capable of staying submerged for about 20 seconds. A mature hippo can stay underwater for as long as six minutes. A mature Common Hippopotamus can grow to 2.7 m in length and weigh a hefty 1.5 tonnes.

The hippo's inner layer of skin is up to 3.5 cm thick and is very strong, but its outer layer of skin is very thin, dries out easily and is sensitive to insect bites. Despite having special mucous-producing glands, the hippo must still moisten its skin regularly with water to prevent dryness and cracking.

experience under his belt, cheekily answered, 'I am not sure if I can tell from its ears or its nostrils!' Graham took me down to see the baby for myself and I was to learn then that hippo calves are often born underwater and this brand-new arrival, only just visible beside the huge bulk of its mother, was peering above the water occasionally to take a quick breath and shake the water out of its tiny ears. It was certainly impossible to determine its gender at this early stage.

Lindy and Billy, the parent hippos, and their calf, which looked to me like a rubber toy in a bathtub and was named 'Happy' by the keepers, were ultimately transferred to Western Plains Zoo where they revelled in an expansive lake with a sandy beach.

These rather comical-looking, seemingly lethargic animals, often remembered for wearing tutus in Disney's *Fantasia*, are actually one of the most active of creatures and are fast walkers, strong and graceful swimmers and ferocious fighters. The wide, open spaces and lakes of Western Plains Zoo suit them well. The much smaller Pygmy Hippopotamus (*Hexaprotodon liberiensis*) is now representing the 'river horse' in the African Waterhole at Taronga Zoo. This animal weighs a more diminutive 275 kilograms and only grows to a length of 1.6 metres.

TARONGA'S SITE AND probably that of many urban zoos, well suits the display of the more active, smaller species in the animal kingdom. With the transfer to Dubbo of the larger and more traditional zoo animals, it was time for Taronga to highlight some of these lesser-known, non-traditional zoo creatures and to introduce visitors to a whole new delightful world of exotic wildlife such as Meerkats, Fennec Foxes, Oriental Small-clawed Otters and the like.

The industrious Meerkats (*Suricata suricatta*), which took up residence at Taronga in 1991, became an instant success and are now one of the Zoo's most-loved animals. These small desert-dwelling mammals from Southern Africa are only about 35 centimetres tall and barely weigh 975 grams. These expert diggers have a highly developed social structure, which, among other behaviours, sees a Meerkat troupe post a guard at a nearby high point to be on the ever-vigilant lookout for airborne predators. This sentry ensures that the rest of the Meerkats can safely forage for food and dig intricate burrows until it's time for the changing of the guard that enables the vigilant ones to feed as well. Zoo visitors, both the young and not so young, love watching these animals furiously digging for food and scurrying about. In winter they can often be seen basking in the heat of specially placed infrared lamps.

Meerkats, with their tiny snouts, paws and bodies continually covered in dirt, often made me shudder to think what life must be like spent with sand constantly in your cossie! Meerkats always make me happy too. Every time I look at them, either the real thing or a photograph, I smile. If my Zoo day was a tough one I would make sure I found time to call by the Meerkat Desert and I could guarantee I would feel instantly better.

The Zoo's new Meerkat Desert exhibit was generously funded by Zoo Friends and was officially opened in April 1991 by their President, Janet Rowe. For this event I invited Tony Rae the captain of Taronga's local rugby league football team, the North Sydney 'Bears', to speak on the importance of strong teamwork, which is something the Meerkats and footballers have in common. It was a very enjoyable school holiday event with the footballers, the Zoo crowds and the Meerkats all obviously having lots of fun. On the same day, we invited another 'Bears' team member, Mario Fenech, to open the nearby Fennec Fox exhibit. It was good to see the very positive impact these exquisite little Meerkats and Fennecs had on the big and macho footballers.

One aspect of Meerkat teamwork, which the footballers probably don't have in common, is the role 'aunties' play in raising the young Meerkats. I was desperate for the Meerkats to

> ## ON GUARD
>
> EVEN IN THE SAFE Zoo environment, Meerkat sentries act as lookouts for aerial predators such as hawks. At Taronga, the airships which motor across Sydney's air space, advertising a variety of messages and giving joy rides over the harbour, particularly send the resident Meerkats running. Sharp barks or growls denote a need for urgent action and the Meerkats dive for cover.

have babies as I could imagine writing a news release titled 'The Meerest of Kats'! Finally, after several years of waiting, the first tiny Meerkats were born and their media debut in March 1998 was a wonderful event. It was the first time Meerkats had been born at Taronga and there was great excitement in the air as the media vied for the best shots of the tiny new arrivals.

There was also a great deal of competition among the new aunties who all were determined to look after the Meerkat babies. It seems they were all furiously vying for the title of 'favourite aunt'. So much so, that the poor Meerkat babies often found themselves in the middle of a 'tug-of-love'. This competition developed to the point where one unfortunate newborn animal had the end of its tiny tail bitten off by an over-zealous auntie.

THE MEERKATS ARE street wise and cunning little creatures. They always provide an enormous challenge and a substantial amount of exercise for their keepers when these fast and tricky animals need to be caught up for their annual vaccinations. I thought this might be a good photo opportunity so I invited a newspaper photographer in early one morning to record this fine example of the Zoo's preventive medicine program in full swing.

Each Meerkat needs to be separated and individually caught, checked and vaccinated. There is a complicated network of small tunnels and nest boxes behind the animals' exhibit area, which are designed as perfect sleeping quarters for the subterranean Meerkats. However, they also pose an enormous challenge when trying to pin down your slippery little Meerkat.

Meerkat Meals

TARONGA'S 11 MEERKATS enjoy four feeds each day. Sunday to Friday breakfasts consist of a juicy mouse or chick and Saturday's breakfast is a raw egg (they particularly love the yolks). Around mid-morning the Meerkats share 900 g of omnivore mix which consists of sweet fruit and sweet potato, a sprinkle of fly pupae and multi-vitamin powder. The keepers finely chop this into 1-cm pieces. In the afternoon the Meerkats share 20 g of mealworms and 22 pieces of dog kibble. The last feed of the day is a sprinkling of live crickets and cockroaches, which are specially bred in the Zoo's insect house.

Young and agile keepers like Louise Ginman and Justine Powell face this challenge annually. They play a game which looks a lot like the thimble and the pea as the keepers and the Meerkats run from nest box to nest box, and tunnel to tunnel, each trying to outsmart the other. Ultimately, the keepers succeed and the indignant Meerkats receive their essential health check and necessary protection against disease. I am sure this extra exercise is good for photographers and zookeepers, too.

NUMEROUS CAMEL ESCAPADES over the years have been recorded in my diaries but three special Zoo moments involving these strange 'ships of the desert' particularly come to mind. The first involved a case of innocent mistaken identity between Fred the office cleaner and Fred a handsome young male Dromedary (*Camelus dromedarius*).

Fred the camel was one of the local identities when I first went to work at Taronga. He and Wendy the camel lived in a yard where Taronga's reptile and amphibian exhibit 'Serpentaria' now stands. Fred began getting a bit too frisky and was becoming a little difficult for the keepers to manage. Camels can be very dangerous animals.

Fred the cleaner was, however, a quiet and gentle man who kept the Zoo offices polished to within an inch of their lives. But our lives were not worth living if we dared to untidy our 'enclosures' or spill anything on Fred the cleaner's newly washed or vacuumed floors in the Zoo administration building.

I, and other members of the Zoo office staff, had sometimes expressed our fears of one day being the subject of Fred's wrath, should we dare neglect to tidy our office or wipe up a spill.

> **REAL CAMELS DON'T EAT QUICHE**
>
> IN THE WILD THE CAMEL feeds on a huge variety of plants including salty and thorny bushes. It also scavenges on bones and dried-out carcasses. It can shed up to 40 per cent of its body weight when food and water are scarce.

For one awful moment one day, I thought that Dave Cody must have been the subject of Fred's displeasure but had gone much too far in retaliation.

Dave came into my office early one morning, looking a bit dishevelled, which was not at all like him. He even had blood splattered all over the front of his khaki shirt. I asked him what had happened and nearly fainted when Dave replied: 'Fred was getting too bossy so I castrated him and it was a bit messy!' I nervously and quickly asked a few more questions, and realised he was talking about Fred the camel not Fred the cleaner!

I always appreciated Zoo Director Jack Throp's wonderful imagination. Each year, apart from dreaming up hundreds of other creative ideas, Jack devised an innovative and unique way to celebrate the end of the Zoo year at the Board's annual Christmas Board luncheon. One of Jack's earliest efforts after he joined the Zoo in 1980 was to stage a living nativity scene. I hired some great costumes and rounded up the appropriate-looking Zoo staff to play the parts of Mary, Joseph, the Three Wise Men and the shepherds. An old music rotunda was decorated to look like a stable complete with a baby doll in a manger and Taronga's best-mannered goats, sheep, donkey and Wendy the camel were chosen and specially trained to add our unique Zoo reality to the nativity scene.

As our nativity scene needed to be in place from around 10.00 a.m. to 4.00 p.m., I suggested that Mary (played by graphic designer Marina Bishop), Joseph (none other than accountant Peter Belcher), the three Wise Men (keepers Tony Carrick, Jonathon Brown and Paul Kirk) and shepherd boy (Willie Soden, son of Zoo Maintenance Manager, Bill Soden) hide some food and drinks in the stable to sustain them throughout what was sure to be a long, hot day.

Although I always understood camels were designed to survive harsh desert conditions and could control the effects of dehydration by minimising water loss because of their kidney action, apparently our Wendy was different or had not read that particular camel handbook. Unfortunately, she became both hungry and thirsty and just as the Board Members arrived to admire our wonderful living tableau, Wendy was seen devouring the Wise Men's rather delicious-looking spinach quiche and grabbing at their Thermos flask, which they had hidden in the manger.

Ghan, the 'citronella camel', was a regular and more recent visitor to the public relations office in the late 1990s. I always knew he was in the garden as the smell of his citronella fly repellent, not the aroma usually associated with camels, wafted in my office window. Fred and Wendy camel certainly never smelled as pleasant.

Keeper Joe Haddock was training young Ghan to walk around the Zoo grounds on a halter and lead rope so visitors could meet him up close. To ensure they didn't receive an accidental kick from Ghan's fly-swatting hooves, Joe smothered the camel with sweet-smelling citronella to keep the insects away. Joe and his pet camel were always welcome in our part of the Zoo and everyone looked forward to their frequent visits. Ghan's soft baby camel hair always felt and smelled wonderful.

Ghan grew and grew and became a very large camel and sadly his visits to the Public Relations office garden stopped. I don't think he could fit through our gate anymore!

ALTHOUGH I CAN STILL fit through the entrance gate, the daily visits to my favourite Zoos have now come to an end for me. My decision to retire was not taken lightly and I did so with a certain sadness. But it was time for a change of lifestyle for me, a time to relax a little and do other things without the time constraints I always seemed to wrestle with over the past 25 years.

I have immense appreciation for the privileged position of trust I held and I will be forever grateful for the opportunities my Zoo life afforded me. I believe I enjoyed the best of Zoo times.

I am still involved with Taronga and Western Plains Zoos but in a much more minor role these days and I continue to look forward to watching the Zoos' exciting plans unfold in the ensuing years. In fact, I see the visual changes even more vividly now every time I visit.

My memories of my privileged Zoo life are like picture postcards in a family photograph album and will stay with me forever.

ACKNOWLEDGMENTS

In compiling these 'Postcards from the Zoo' I have been willingly assisted by numerous and very encouraging friends and colleagues. I now warmly thank these extremely generous people who have contributed their own personal thoughts and recollections, special memories, personal photographs, editorial expertise, publishing experience, legal and IT advice, specialist contacts and all essential animal facts and figures:

The Australian Society of Authors
Elizabeth Brown
Sara Brice
Phil Cameron
Christopher Cheng
Marianne Cochrane
David Cody, BEM, AO
Guy Cooper
Heather Dengate
Karen Edwards
Kevin Evans
Simon Finlay

Acknowledgments

Dr Ted Finnie
Maria Finnigan
Elizabeth Gibbons (née George)
Louise Ginman
Paula Goodyer
Elizabeth Hall
HarperCollins's team — Associate Publisher Helen Littleton, Tegan Murray and Gwenda Jarred
Paul Hayton
Libby Kartzoff
Dr Eisuke Kashima
Davin Kroeger
John Lemon
Sally Loane
Steve McAuley
Margaret Miller
Justine Powell
Chris Riley
Bruce Robertson
Ben Rushton (Macaw photograph)
Roy and Doreen Scrivener
Dynah Skillman
Dale Stead
Dr Larry Vogelnest
Phil Whalen

I should make particular mention of my kind and gentle editor, Kate Pollard of HarperCollins, and thank her most especially for her guidance, enthusiasm and expertise.

I must especially and very fondly thank my husband, Robert, for his love, humour, encouragement and patience always and for his unwavering support of and interest in 'all things zoo'.

In compiling these vignettes I have been given generous access to the Taronga Zoo Archive, the Zoo Education Centre library and the Zoo websites *www.zoo.nsw.gov.au* and *www.zootopia.org.au*. I have also referred to, been inspired by or have quoted from the following references:

Allin, Michael, 1999, *Zarafa*, Hodder Headline, London.
Burnie, David (ed.), 2001, *Animal*, Dorling Kindersley Ltd, London.
Graves, Eleanor (ed.), 1976, *Elephants & Other Land Giants*, Time-Life Films, New York.
IUCN Red List of Threatened Animals, International Union for Conservation of Nature and Natural Resources.
Matthews, L. Harrison, 1971, *The Life of Mammals*, Vol 2, Universe Books, New York.
Pizzey, Graham, 2001, *Australian Bird-Garden, The*, Angus & Robertson, Sydney.
Strahan, Ronald (ed.), 1983, *The Australian Museum Complete Book of Australian Mammals*, Angus & Robertson, Sydney.
Strahan, Ronald, 1991, *Beauty and the Beasts: A History of Taronga Zoo, Western Plains Zoo and Their Antecedents*, Surrey Beatty, Chipping Norton.
Woodford, James, 2001, *The Secret Life of Wombats*, Text Publishing, Melbourne.

Darill Clements

In 1975 Darill Clements answered an advertisement for a public relations assistant at Sydney's world-renowned Taronga Zoo. She had no animal background but immediately embraced Zoo 'culture' and became besotted with animals, which enriched her life immeasurably.

Darill's commitment to the Zoological Parks Board's two great Zoos and to wildlife conservation over a quarter of a century is undisputed. She demonstrated a refreshing joy in being part of the achievements of New South Wales' two Zoos during a process of continuous evolution.

She contributed to many wonderful projects in Zoo history, playing a key creative, media communication and organisational role in hundreds of events. Some highlights include the opening of Western Plains Zoo at Dubbo, the visit by the Giant Pandas, the collection and transportation of Black Rhinoceros from Zimbabwe to Australia, the transfer of koalas from Australia to Japanese zoos and the establishment and maintenance of a sister Zoo relationship with Nagoya Higashiyama Zoo, and the return of zoo-bred Prezwalski's Horses to the wilds of Mongolia. Darill retired from the Zoos in December 2000.

In the Queen's Birthday Honours in June 2001, Darill was awarded the Public Service Medal for her outstanding service to New South Wales's two Zoos.

Rick Stevens

Multi-award-winning photographer Rick Stevens joined the *Sydney Morning Herald* in 1963 and began photographing the animals of Taronga and Western Plains Zoos in the mid–1970s.

His interest in wildlife and the environment stretches back to a childhood spent in Zimbabwe, Africa, and despite warnings to avoid taking photographs of animals and children, Rick finds animals great subjects to deal with. They are much less difficult than most people and he says they never complain that he has taken their bad side.

Rick admires greatly the dedication of the Taronga and Western Plains Zoos staff and enjoys working with them. In fact, he says his photography is always a joy, not work. Photographing the animals at the Zoos is an honour and has given him, and the millions of *Herald* readers over the years, an opportunity to learn more, and to appreciate their magnetic beauty and their place in the environment.